COPING WITH CLIMATE VARIABILITY

This book is dedicated to
Robert Kingamkono
(1961–2001)

Coping with Climate Variability

The Use of Seasonal Climate Forecasts in Southern Africa

Edited by
KAREN O'BRIEN
CICERO, University of Oslo, Norway
COLEEN VOGEL
University of Witwatersrand, South Africa

LONDON AND NEW YORK

First published 2003 by Ashgate Publishing

Reissued 2018 by Routledge
2 Park Square, Milton Park, Abingdon, Oxon OX14 4RN
711 Third Avenue, New York, NY 10017, USA

Routledge is an imprint of the Taylor & Francis Group, an informa business

Publisher's Note
The publisher has gone to great lengths to ensure the quality of this reprint but points out that some imperfections in the original copies may be apparent.

Disclaimer
The publisher has made every effort to trace copyright holders and welcomes correspondence from those they have been unable to contact.

A Library of Congress record exists under LC control number: 2002110616

ISBN 13: 978-1-138-70756-6 (hbk)
ISBN 13: 978-1-138-70749-8 (pbk)
ISBN 13: 978-1-315-19994-8 (ebk)

Contents

List of Figures

List of Tables

List of Contributors

Channing Arndt is an Assistant Professor in the Department of Agricultural Economics at Purdue University and a Faculty Associate with the Center for Global Trade Analysis.

Melanie Bacou is an Analyst and Web Specialist with the Center for Global Trade Analysis at Purdue University.

Anna Bartman is a specialist scientist at the South African Weather Service, Pretoria, involved in seasonal forecasting research.

Roger Blench is a Research Fellow of the Rural Policy and Environment Group at the Overseas Development Institute in London. He has published extensively on issues relating policy to the natural environment.

Louise Bohn is a Senior Research Associate at the Climatic Research Unit, University of East Anglia, United Kingdom.

Antonio Cruz is a Ph.D. candidate in the Department of Agricultural Economics at Purdue University. He is on leave from his duties in the Ministry of Planning and Finance, Mozambique.

Hamisi Dihenga is a Senior Lecturer in the Department of Agricultural Engineering and Land Planning at Sokoine University of Agriculture in Morogoro, Tanzania.

Maxx Dilley is a Research Scientist in disaster and risk management at the Columbia University International Research Institute for Climate Prediction.

Mike Harrison was a manager of the first SARCOF series and has subsequently extended the work of delivering seasonal forecasts to end users through management of the international WMO CLIPS Project. He is now at the UK Met Office and is engaged in further research into the interaction between climate, the environment, and society.

Jerry Hudson is a Ph.D. candidate in the Sociology Department at Colorado State University in the United States.

Nganga Kihupi is a Senior Lecturer in the Department of Agricultural Engineering and Land Planning at Sokoine University of Agriculture in Morogoro, Tanzania.

Margaret Kingamkono is a Research Officer in the Ministry of Agriculture and Food at Selian Agricultural Research Institute, Tanzania.

The late *Robert Kingamkono* was an Assistant Lecturer in the Department of Physical Sciences at Sokoine University of Agriculture in Morogoro, Tanzania.

Emsie Klopper is a specialist scientist at the South African Weather Service, Pretoria, involved in seasonal forecasting research. She has recently completed her Ph.D. on the use of seasonal forecasts in commercial farming sector.

Jolamu Nkhokwe is a scientist in the Malawi Meteorological Department, with expertise in climate forecast production and their applications.

Karen O'Brien is a senior research fellow at CICERO (Center for Climate and Environment Research – Oslo) in Norway. Her research focuses on climate impacts and vulnerability to global change.

Jennifer Phillips is an assistant professor at the Bard Center for Environmental Policy at Bard College in Annandale, New York, and an adjunct researcher at the International Research Institute for Climate Prediction. Her primary research interest is in improving the communication of climate information in farming systems.

Winifrida Rwamugira is a Senior Lecturer in the Department of Crop Science and Production at the Sokoine University of Agriculture in Morogoro, Tanzania.

Anne Thomson is a senior economist at Oxford Policy Management, working in the Rural Development and Natural Resource Area

Coleen Vogel is an associate professor of geography at the University of the Witwatersrand, South Africa. Her research focuses on the human dimensions of global environmental change, with a particular emphasis on droughts and vulnerability to change in the southern African region.

Neil Ward is Head of Forecast Development at the International Research Institute for Climate Prediction, Lamont-Doherty Earth Observatory of Columbia University, New York, U.S.A.

Preface

MIKE HARRISON

Recently a new scientific tool has emerged that promises to further reduce the risks to agriculture and food security across southern Africa: seasonal climate prediction. The idea is to supply advanced knowledge of weather through the growing season, and provide it in time to enable farmers to make decisions on optimum strategies to maximize each season's harvest. This idea has become reality through advances in observations of the atmosphere and, particularly, of the oceans, through improved insights into the workings of the atmosphere-ocean system, and through the development of new models, some simple, some requiring the largest of computers, to link the observations and insights and to provide the predictions. The new technology faced its first major scrutiny during the El Niño event of 1997/98 after having been presented directly to the user community through the sequence of Southern African Regional Climate Outlook Forums (SARCOF) that began in September 1997. Although seasonal predictions promise a substantial beneficial return on costs involved, their implementation also presents a challenge to both the scientific and user communities.

The challenges range from scientific/technological to institutional and social. Scientifically, seasonal predictions face the challenge of chaos. Chaos is a physical reality that limits the amount of predictability available in many environmental systems, not least seasonal climate forecasting. While much of southern Africa tends to experience reduced rainfall, perhaps even drought, when El Niño events occur, a prediction that an El Niño will occur is not sufficient to guarantee that the region will become contemporaneously drought stricken; the 1997/98 experience was a clear case in point. Unfortunately not all proponents of this nascent science provided their forecasts during 1997 with due levels of caution, and the credibility of the discipline was dealt a partial blow.

The practical outcome of chaos as seen by users is that forecasts are presented as probabilities and at temporal and spatial scales rather larger than those with which the client community is familiar with from daily weather predictions. Users have had difficulties in coming to terms with predictions stated in the form of: "there is a 55% probability that total rainfall over an area of 200km x 200km through the months December to February will be in the top tercile." There is no doubt of the scientific rectitude of this statement, but there is equally no doubt of the opaqueness of this statement from the user perspective. Users have become familiar with scientists presenting forecasts in terms they understand; thus with seasonal predictions they naturally expect to be reliably told what is going to happen through the season, and at temporal and spatial scales appropriate to their planning needs. Scientists, aware of the limitations imposed by chaos, have attempted to hold a legitimate line.

The institutional issues relate to the limited number of facilities, mainly in the developed world, that are capable of undertaking the necessary monitoring and prediction activities. Forums of the SARCOF type, now regular features of the calendar in many parts of the world, have been part of the response. Other activities have included capacity building, the development of simpler regional prediction models, and the transfer of more complex models where facilities permit. Efforts have been made in several instances not only to develop collaborations that lead to the output of forecasts on the scales needed, but also to deliver the technology to make those predictions locally.

The societal issues are legion. Education of the end users and of the intermediaries who deliver the information to the end users, is one prime aspect. Distribution methods themselves are a second aspect. Confidence development is perhaps the key, as users will resist incorporating information that they do not fully believe into decision processes, particularly if doing so may have negative, and perhaps severe and relatively long-term, consequences. Why, it might be asked, should agriculturalists, who in the past have come to terms with a series of risks, now introduce a new risk before the benefits are adequately demonstrated?

Agriculturalists in southern Africa have come to terms with risk over many centuries. The experience gained has been used to minimize the numerous threats to food supplies offered by climate variability, disease, pestilence and hostile neighbors. Coping strategies have included diversification, irrigation, water storage, and, *in extremis*, appropriation or even migration. Forecasting of the seasonal rainfall was, and still is, called upon by many traditional societies as an additional aid to help manage risks, with both phenology and the gods used to assist in the decision making process. Over time, societies became dependent on this cumulative experience and in general achieved management of food supplies sufficient to satisfy the needs of the population. Modern technologies have further reduced risks, with mechanization, fertilization, cultivarization and commercialization all leading to improved and more reliable yields. Even so, these advances are not available to all southern African farmers and they still cannot guard against all risks–especially those related to climate variability. In practice commercial farmers, with their relatively high levels of resilience, have tended to adopt and gain the benefits of seasonal prediction more readily than subsistence farmers.

Southern Africa has become one of the most important regions of the globe in which the development of applications of seasonal forecasts is being piloted. Rainfall variability is critical to agricultural activities in the region and there is a relatively high degree of predictability of that variability on seasonal time scales, in part because of links with the El Niño phenomenon. Southern Africa, as mentioned above, was the first region to host a Regional Climate Outlook Forum, and scientists in the region have proceeded to develop the Forum process to engage climatologists, intermediaries, end users, and policy makers. Southern Africa has a wide range of agricultural activities, much of it only rain fed, from commercial to subsistence, covering a plethora of crops and animals, and undertaken in a variety of climatic regimes. Southern Africa also has a rich and varied cultural history that has led to a multiplicity of agricultural practices. Long-term projections suggest that ensuring food security in the region may become more complex over the next century, as the climate becomes progressively drier.

The papers in this book cover many of the issues and experiences of developing the application of seasonal prediction in southern Africa. Among the authors are some of the leading pioneers in this new scientific art. Their work outlines clearly the difficulties involved, and in some cases provides viable resolutions. More solutions are still required but the basis built through this work will provide a vital foundation on which these solutions will be sought. There can be little doubt that delivering those solutions will require the synergetic engagement of everyone, from scientists to end users; neither community can overcome the difficulties alone. The scientists will strive continually to improve their capabilities and to tie them in more closely with the needs of the users; at the same time the users will need to evaluate their approaches in order to determine how the benefits of existing capabilities might be optimally utilized. The processes outlined in this book are only a beginning, but a very important beginning. There are substantial benefits on offer.

Acknowledgments

We would like to express our thanks to the people who have helped us to prepare and complete this book. We are particularly grateful to Arne Dalfelt at the World Bank for supporting the initiative to convene the 1999 Workshop on User Responses to Seasonal Climate Forecasts in Southern Africa in Dar es Salaam, Tanzania to discuss user responses to seasonal forecasts in southern Africa. We would also like to acknowledge the contributions of others who participated in this workshop, including Ben Hochobeb, Tharsis Hyera, Amin Bakari Iddi, Maynard Lugenja, and Lars Otto Naess. We would like to extend thanks to Wendy Job, Lynn P. Nygaard, Margot Rubin, Santiago Olmos, and Tone Veiby for editorial assistance with the manuscript, and to Carolyn Court and the editors at Ashgate Publishing for their kind understanding and patience regarding the many delays on this book. And finally, we would like to thank our families for coping with us throughout this project.

It was with great sadness that we learned of the death of Robert Kingamkono, a key researcher and collaborator on studies of user responses to seasonal forecasts at Sokoine University of Agriculture in Tanzania. We would like to dedicate this book to him.

PART I:
SEASONAL CLIMATE
FORECASTS

1 Climate Forecasts in Southern Africa

COLEEN VOGEL AND KAREN O'BRIEN

> Seasonal forecasts are a means of helping users make informed
> decisions related to climate variability. The users of seasonal
> climate forecasts are an essential part of the forecasting process,
> both as users of the information and as sources of feedback to the
> forecasters and the scientific community.
>
> (IRI, 2001a, p. 156)

Introduction

The weather and climate over southern Africa is highly variable, with
extreme events such as droughts and floods frequent occurrences. Coupled
to the vagaries of climate and weather are other human problems related to
conflicts, increasing HIV/AIDS, structural adjustment, and foreign debt.
For those trying to make a living in the agricultural sector, climate
variability adds to the complexity of decision-making (Benson and Clay,
1998). Long-term climate change resulting from increasing atmospheric
concentrations of greenhouse gases may, furthermore, influence climate
variability, changing both the frequency and magnitude of extreme events
(McCarthy et al., 2001).

Projected impacts of climate change based on models and other studies
include reductions in potential crop yields in most sub-tropical and tropical
regions, decreased water availability in the sub-tropics, and increases in
droughts and other extreme events (McCarthy et al., 2001). Climate
variability and long-term climate change thus pose serious challenges for
southern Africa, requiring concerted efforts in mitigation and adaptation, as
well as an improved ability to live with change (Hulme, 1990; Sharma et

al., 2001). One way of increasing capacity to cope with climate variability is through reliable and accessible seasonal climate forecasts. Seasonal climate forecasts have been developed to assist in decision making and improve the ability of users to live with climate risk, most notably fluctuations in rainfall.[1]

There are two principal sources contributing to rainfall variations in southern Africa. The first is the El Niño Southern Oscillation (ENSO) phenomenon and the second is sea surface temperature variability in the South Atlantic and Indian oceans (Mason and Jury, 1997; Lindesay, 1998). Because of a strong connection between sea surface temperatures, large-scale near surface pressure systems, and rainfall, scientists have been able to use the relationship between ENSO and rainfall as inputs to seasonal climate forecasts (for overviews, see Philander, 1990; Battisti and Sarachik, 1995; Mason, 1997; Mason and Jury, 1997; Cane, 2000; Ropelewski and Foland, 2000; Goddard et al., 2001). Research has indeed shown that ENSO events can be predicted with some skill and enough lead time to potentially contribute to cost-saving decisions in various sectors (Cane, Eshel and Buckland, 1994; GCOS, 1995). Improved climate information in the form of forecasts can be included together with existing management tools, such as early warning systems and vulnerability mapping, to help reduce risks associated with climate variability.

This book is concerned with seasonal forecasts and their potential contributions to agriculture and food security in southern Africa. Over the past several years there has been mounting interest, in southern Africa and elsewhere, in improving forecast skill and expanding the dissemination and use of seasonal forecasts, with the objective of maximizing their value to the user-community (see Gibberd et al., 1995; Glantz, Betsill, and Crandall, 1997; Thomson, Jenden, and Clay, 1998; Orlove and Tosteson, 1999; Dilley, 2000; Eakin, 2000; Glantz, 2000). Yet despite increased success in producing forecasts, much still needs to be done to assess and improve their value to a variety of users. Where useful, forecasts need to be made more accessible and understandable. This is particularly true in southern Africa, where the potential benefits of forecasts are great, but where the challenges to further utilization of forecasts are even greater.

The objective of this book therefore is to bring together some of the recent research on user-responses to seasonal forecasts in southern Africa, drawing out key issues and the greatest challenges that remain. The topics covered in this book include the appropriateness of forecasts as an agricultural decision-making tool for a variety of farmers and end-users; the dissemination, uptake, and use of forecasts, and finally, the degree of "fit"(including organizational fit) between users of forecasts and producers of forecasts (e.g., Orlove and Tosteson, 1999). The chapters included in this book are linked to case studies presented at the 1999 Workshop on User Responses to Seasonal Climate Forecasts in Southern Africa,[2] as well as contributions from participants in the Southern Africa Regional Climate Outlook Forum (SARCOF). The book also draws upon previous studies on the uptake and use of climate forecasts in southern Africa by end-users (e.g., Klopper, 1999; NOAA-OGP, 1999; Buizer, Foster, and Lund, 2000; Vogel, 2000; O'Brien et al., 2000; Archer, 2001; Phillips, Makaudza and Unganai, 2001, Walker et al., 2001; Ziervogel, 2001). By synthesizing research on user responses to seasonal forecasts in southern Africa, we hope to contribute to a wider dialogue on the use of climate information, both within and across different regions and sectors.

We begin, in this chapter, by describing the southern Africa region and providing some background on climate and climate variability, including the role of ENSO as one of the key influences of rainfall variability in the region. The production of seasonal climate forecasts for southern Africa is discussed, followed by their use, uptake, and overall potential application in the region. We then identify what we see as the major challenges to the expanded use of forecasts in southern Africa. These challenges emerge from the research presented in subsequent chapters, which are summarized briefly below.

Southern Africa: The Context for Seasonal Forecasts

The spatial context for this book is southern Africa, a region broadly considered to include fourteen members of the Southern African Development Community (SADC):[3] Angola, Botswana, Democratic Republic of Congo, Lesotho, Malawi, Mauritius, Mozambique, Namibia, Seychelles, South Africa, Swaziland, Tanzania, Zambia, and Zimbabwe (Figure 1.1). The region is characterized by both geographic and demographic diversity. The total estimated population of the SADC region was approximately 195 million in 2000 (see Table 1.1). Among the fourteen SADC countries, the recently-included Republic of Congo has the largest population, with approximately 50.5 million people. This is followed by South Africa (44.5 million), Tanzania (30.8 million) and Mozambique (16.9 million). By contrast, countries such as Namibia and Botswana have relatively small populations (1.8 and 1.6 million, respectively).

Figure 1.1   Map of the SADC region (excluding Mauritius and Seychelles)

The distribution of land area among the SADC countries varies greatly. The Republic of Congo alone constitutes about 25% of the region and together with Angola and South Africa includes more than 50% of the total area. The share of land classified as arable also varies within the region. For example, 10% of South Africa's land is considered arable, whereas only 3% of Tanzania's land area is classified as such, compared with a mere 1%

in Namibia. While countries such as Angola and Zambia benefit from a large agricultural resource base relative to the size of their populations, Malawi, Lesotho, Botswana, Swaziland, and Namibia all show signs of acute pressure on available agricultural production resources. Furthermore, there are growing tensions and conflicts over land, as is the case in Zimbabwe.

Table 1.1 Statistical data for the SADC region

Country	Total area ('000 sq km)	Population (million, 1999)	GDP per capita (in PPP US$)
Angola	1 247 000	12.8	3179
Botswana	585 000	1.5	6 872
D.R.C.	2 435 409	49.6	801
Lesotho	30 355	2.0	1 854
Malawi	118 484	11.0	586
Mauritius	1 968	1.2	9 107
Mozambique	799 380	17.8	861
Namibia	824 269	1.7	5 468
Seychelles	455	0.8	7 700
South Africa	1 219 090	42.8	8 908
Swaziland	17 364	0.9	3 987
Tanzania	947 532	34.3	501
Zambia	752 614	10.2	756
Zimbabwe	391 109	12.4	2 876
SADC	9 369 620	199	

Source: United Nations Development Programme, 2000

In South Africa and Zambia, more than half of the population lives in urban areas, whereas in most of the other countries only about 25–35% of the population is considered urban. Nevertheless, the urban population of SADC countries has been increasing rapidly, and rural-urban migration is expected to continue growing in the years to come. The consequences of this urbanization for the agricultural sector are many, as it influences supply and demand of labor and food, and other factors of production. The southern Africa region is also experiencing an array of impacts related to the spread of HIV/AIDS. While there are problems related to the accuracy of statistics, indications are that the prevalence and infection rate of HIV as a percentage of the adult population is highest in the SADC region, as compared to other areas on the African continent. South Africa, where an estimated 4.2 million are infected with HIV, is said to have more people living with HIV than any other country (Dorrington et al., 2001). The implications of HIV for economic growth and for households actively involved in agriculture are difficult to quantify. In fact, many of the impacts are only beginning to emerge, and thus are not yet well documented. Nevertheless, the spread of HIV/AIDS, coupled to other health problems, including malaria and cholera, is resulting in enormous costs to society, both in urban and rural areas (Topouzis and du Guerny, 1999).

The regional economy of southern Africa is dominated by the contribution of South Africa and Botswana, which account for more than three-quarters of regional gross domestic product (GDP). Although the economic structure of the 14 countries is diverse in terms of human and natural resources, there are a number of important similarities. Historically, many of these countries have experienced the same problems and challenges, including colonialism, wars, political instability, drought, and economic crises leading to fluctuations in export commodity prices and declines in the capacity to import. Income inequalities are high within each country, and development is unevenly distributed. As a consequence of economic instability, many of the countries in southern Africa were obliged to adopt severely deflationary structural adjustment programs in the 1980s. Aimed at controlling inflation, eliminating current accounts deficits, and alleviating balance of payments problems, these structural adjustment programs have had uneven impacts on African economies (World Bank, 2000; IFAD, 2001). The impacts on the agricultural sector have been particularly ambiguous; efforts to liberalize and privatize agricultural markets have benefited some farmers, while marginalizing others (Leichenko and O'Brien, 2002).

Agriculture is one of the key sectors for potential and actual use of climate information. In fact, agriculture forms the mainstay of many of the economies in southern Africa (Table 1.2), and plays a particularly important role in terms of rural livelihoods (Whiteside, 1998). For farmers in southern Africa, rainfall is a critical factor. When the rains fail, farm output decreases. Too much rain, on the other hand, can result in flooding that makes harvesting and marketing difficult. Climate variability can thus have strong impacts on the livelihoods of farmers, and on national economies. Although climate variability is just one of a suite of factors that influence end-user's decision environment, seasonal climate forecasts present a potential tool that could assist users in better managing their risks.

Table 1.2 Agriculture in the SADC region

Country	Importance of agriculture to economy (% GDP, 1996/97)	Labor force in agriculture (% of total, 2000)	Irrigated land (% of arable land, 1994)
Angola	13	72	2.5
Botswana	4	45	0.3
D.R.C	59	63	0.2
Lesotho	14	38	0.3
Malawi	45	83	1.3
Mauritius	8	12	20.0
Mozambique	35	81	2.7
Namibia	11	41	0.9
Seychelles	4	-	-
South Africa	5	10	10.2
Swaziland	10	34	39.3
Tanzania	56	80	4.0
Zambia	23	69	0.9
Zimbabwe	28	63	3.6

Source: FAO Agricultural Data, 2000

Climate, Climate Variability, and the Seasonal Forecasts

Having provided a brief description of the socio-economic context for climate forecasts in southern Africa, attention turns to climate controls in the region and the production and use of seasonal climate forecasts. Variability is an inherent characteristic of the climate of southern Africa. The climate of the region can be defined as predominantly semi-arid, with high intra- and inter-annual rainfall variability (Lindesay, 1998; Tyson and Preston-Whyte, 2000). Average annual rainfall in southern Africa is just under 700 mm. Rainfall is generally highest in tropical Africa, towards the equator, and decreases to the south and west. Large spatial variations prevail, with some desert areas receiving less than 200 mm and some highland areas receiving over 2000 mm (Lindesay, 1998; Garanganga, 1998). Furthermore, local exceptions occur within seemingly homogenous climate zones.

Rainfall also varies temporally in the region. Some areas in southern Africa experience one rainy season, while other areas experience two. For example, northern Tanzania experiences both a short rainy season and a long rainy season. Generally, however, the rainy season in southern Africa extends from October/November to April, reaching a peak between December and February. Most of southern Africa receives more than 75% of its mean annual precipitation during the rainy season, and some parts receive as much as 90% in this period. Significant precipitation is unusual after mid-May (Hulme, 1996).

The atmospheric dynamics between tropical and mid-latitude weather systems, as well as convective variability, contribute significantly to rainfall in southern Africa (Lindesay, 1998; Tyson and Preston-Whyte, 2000). Various synoptic controls also influence rainfall, including tropical, mid-latitudinal, and high-pressure circulations (Tyson and Preston-Whyte, 2000). The ITCZ (Inter Tropical Convergence Zone), for example, is a region characterized by much convectional activity resulting in rainfall in several southern African countries, particularly during the summer months. The mixing processes that produce this convection, however, are complicated. The penetration of mid-latitudinal disturbances into the tropics, for example, plays a role in generating tropical rainfall (Lindesay, 1998).

In low latitudes, tropical convective processes are the dominant rainfall producers. In a limited area of tropical East Africa, however, the effects of monsoonal winds also influence rainfall seasonality (Hastenrath, 1985; Lindesay, 1998). Other major circulation types that play an important role in producing rainfall in southern Africa include tropical disturbances in the easterlies, temperate disturbances in the westerlies, cloud bands that link tropical and temperate disturbances, and ENSO (Harrison, 1984, 1986; Lindesay, 1998; Tyson and Preston-Whyte, 2000).

ENSO and Southern African Rainfall

The El Niño Southern Oscillation (ENSO) phenomenon is the climate engine influencing rainfall on an inter-annual time scale. Indeed, ENSO has been described as one of the most important determinants of year-to-year climatic variability and severe impacts around the globe (Gibberd et al., 1995; Glantz, 1996; Cane, 2000). The understanding of ENSO and its use for forecasting has developed over the years (e.g., Neelin et al., 1998).

ENSO refers to a coupled ocean-atmosphere system that links changes in atmospheric pressure and sea-surface temperatures over the Southern Pacific Ocean. El Niño is the term used to describe the extensive warming of the upper ocean in the tropical eastern Pacific. This ocean-atmosphere phenomenon occurs when the air pressure gradient between the central and western parts of the Pacific Ocean weakens, resulting in a dramatic rise in ocean temperatures, usually coupled with an increase in rainfall in the eastern Pacific (Peru) and a decrease in rainfall in the western Pacific (Indonesia and Australia) (Philander, 1990; Allan, Lindesay, and Parker, 1996).

ENSO events usually occur periodically, every two to seven years (Philander, 1990). A typical ENSO cycle lasts for three or more seasons, developing through several phases, from the warming of the oceans to the return to normal temperatures, followed by a cooling known as La Niña. Changes in the location and concentration of atmospheric and oceanic heat associated with ENSO alter atmospheric circulations and lead to changes in climate patterns around the globe. The influence of ENSO events on the range of climate outcomes locally depends on the region's climate, the season, and also the strength and spatial distribution of the ENSO-related SST anomalies (Lindesay, 1998; Goddard et al., 2001). Although the impacts of El Niño are global, most attention has been given to the regional impacts of the phenomenon.

The ENSO phenomenon is one of the major influences on interannual variability over southern Africa (Lindesay, Harrison, and Haffner, 1986; Matarira, 1990; Mason and Jury, 1997). It is estimated that El Niño accounts for between 30 to 35% of climate variability in parts of the region (Makarau and Jury, 1997; Lindesay, 1998). More particularly, ENSO has also been shown to be usually, but not always, linked to droughts in southern Africa (e.g., Lindesay, Harrison and Haffner, 1986; Janowiak, 1988; Mason and Jury, 1997). For example, below-average rainfall and droughts in many parts of the region coincided with a remarkably strong El Niño event in 1982/83 (Lindesay, 1998). Such droughts have significant impacts on regional and local food production, with both economic and social consequences (Benson and Clay, 1998).

Production of Long-term Seasonal Forecasts

The scientific understanding of ENSO events has improved over the past two decades (see Rasmusson and Carpenter, 1982; Philander, 1983, 1990; Ropelewski and Halpert, 1987; Halpert and Ropelewski, 1992; Latif et al., 1998; McPhaden et al., 1998; Neelin et al., 1998; Stockdale et al., 1998; Cane, 2000). The growth in expertise related to ENSO forecasts was spurred on by the impacts of the early 1980s ENSO event and by university-based and international research programs, including the Tropical Oceans Global Atmosphere (TOGA) Program; National Oceanic and Atmospheric Administration (NOAA) Climate Prediction Center, and the International Research Institute for Climate Prediction (IRI) (McPhaden et al., 1998; Orlove and Tosteson, 1999; Mason et al., 2000; Cane, 2000; Buizer, Foster, and Lund, 2000).

Predictions of El Niño have been developed using a variety of models, simulations, statistical methods and empirical methods (Latif et al., 1998; Cane, 2000; Mason and Mimmack, 2002). Traditionally, operational monthly and seasonal climate prediction has been based on empirically derived statistical relationships (Cane, Zebiak, and Dolan, 1986; Zebiak and Cane, 1987; Barnston et al., 1994; Cane, 2000; Ropelewski and Folland, 2000). Efforts include the use of both univariate and multivariate approaches (see Mason, 1997; Stern and Easterling, 1999; Washington and Downing, 1999). Several examples of current empirical climate prediction techniques appear in the NOAA Experimental Long-Lead Bulletin (Climate

Prediction Center, 1992–7; Ropelewski and Folland, 2000). Increasingly, General Circulation Models (GCMs) are also being used to produce operational forecasts or experimental and extended range forecasts.

Sea surface temperatures (SSTs) and their linkages to atmospheric circulation systems are one of the key elements used in developing seasonal climate forecasts (Stern and Easterling, 1999; Goddard et al., 2001). In Peru, Ecuador, Australia, the Pacific Islands, as well as parts of southern Africa, precipitation and temperature are coupled to the ENSO phenomenon (Lindesay, Harrison, and Haffner, 1986; Janowiak, 1988; Jury, 1995; Mason and Jury, 1997). This linkage is usually only well-defined in a few places (Landman and Mason, 2001; Mason and Goddard, 2001). Nevertheless, the skill in predicting climate variations based on SSTs is relatively high, and the future of these forecasts is potentially very promising (Stern and Easterling, 1999). Most of the forecasts referred to in this book are based on SSTs, which provide an indication of possible ENSO-related climate conditions that may be developing in southern Africa.

Despite the improvement in forecasting methods there is still research needed on the level of predictability and reliability of seasonal forecasts. As Cane (2000, p. 42) notes: "There is certainly room for improvement" and progress in the analysis of the limits to ENSO forecast skill is essential. A decrease in skill levels has, for example, been noted in the 1990s (Kirtman and Schopf, 1998; Goddard et al., 2001). ENSO, it must be remembered, is also not the only mode of climate-variability with large-scale impacts (Cane, 2000). There can also be a range of possibilities associated with ENSO depending on the strength of the event, the influence that it has on a particular area, and the complex dynamics and responses of climate to ENSO. The relationship between agriculture and rainfall and temperature associated with ENSO in parts of southern Africa, for example, "is not entirely reliable and straightforward: 1992 was the most severe drought in at least 150 years in southern Africa, but only a moderate El Niño" (Cane, 2000, p. 45). The role of the oceans in modulating the influence of ENSO, particularly for the southern African region, is also an additional consideration. The links between ENSO and SSTs in the Indian Ocean for the Southern African region have emerged as critical factors requiring more research (Tennant, 1996; Landman and Mason, 1999).

Notwithstanding these limitations, ENSO remains an important phenomenon that modulates southern African rainfall. Improved understanding of the relationship between ENSO and southern African rainfall remains a critical challenge, particularly given the possibility of changes in atmospheric circulation associated with long-term climate change. With these cautionary remarks in mind, attention now shifts to current efforts to improve the communication and uptake of seasonal forecasts.

Use of Seasonal Forecasts

The ability to model ocean-atmosphere interactions and thereby predict seasonal-to-interannual climatic variability across broad regions of the globe has improved, and predictive skill is now sufficiently high to enable forecast producers to release information on ENSO-related weather phenomena to the public (Stern and Easterling, 1999). The 1997/98 rainfall season marked the first year that seasonal climate forecasts were widely disseminated in southern Africa. Unlike the ENSO event of the early 1980s, the onset and related potential impacts of the 1997/98 event were anticipated before they occurred. By using data collected from an array of buoys in the Pacific Ocean and the various GCMs and statistical models, and by incorporating recent advances in seasonal-to-interannual climate variability research, it was possible to predict the 1997/98 El Niño event: "The development of the 1997/98 El Niño provided an ideal opportunity to generate climate forecasts on an operational basis" (Mason et al., 2000, p. 1855).

Although seasonal forecasts have been disseminated through a variety of channels over the past decade,[4] a concerted effort to produce and disseminate consensus forecasts was initiated only in the late 1990s, when a series of Climate Outlook Forums were organized by the NOAA's Office of Global Programs (NOAA-OGP). The 1997/98 El Niño event, coupled with ever-growing concerns about potential climatic changes associated with global warming, contributed to the heightened awareness and desire to better manage climate fluctuations (IRI, 2001a, 2001b). Through regional meetings and pilot projects established around the world, the objective was to develop consensus forecasts, and to link international producers of forecasts with actual and potential users of these forecasts (Buizer, Foster, and Lund, 2000).

Included among these were a series of meetings in southern Africa referred to as the Southern African Regional Climate Outlook Forum (SARCOF).[5] SARCOF represents an effort to promote the dissemination of consistent and clear consensus forecasts to the user community, and to minimize the confusion that arises when conflicting forecasts from various sources are heard (NOAA-OGP, 1999). SARCOF thus aims to draw together existing climate products into one seasonal forecast, or Regional Climate Outlook. Another objective of SARCOF and other forums has been to promote regional capacity to produce and apply seasonal forecasts. This involves addressing gaps in training and technical capability, and facilitating research cooperation and exchange of information. Different forecasts are presented and discussed at the SARCOF meetings, and a "consensus" forecast is then produced for the southern African region.[6] Seasonal climate forecasts provide probabilistic estimates of total rainfall relative to a 30-year period. They are different from short-term weather forecasts because they cover relatively larger regions and longer time spans. The resolution of the forecasts is rather coarse; one set of terciles (showing probabilities of below normal, normal, and above normal rainfall) can correspond to a region covering several hundred square kilometers (see Figure 1.2).

The SARCOF forecasts are currently limited to information about total seasonal rainfall. Consequently, the weather at particular points and at specific times may sometimes appear to contradict the climate forecast. Seasonal forecasts say little, for example, about the onset or cessation of rainfall, the anticipated length of the growing season, or the spatial distribution of rainfall (NOAA-OGP, 1999). As illustrated in this book, these factors frustrate the uptake and effective use of forecasts in the SADC region.

Reaching the End Users

Among the users represented at SARCOF meetings are international donor agencies, government ministries, commercial farmers' organizations, and researchers. Many of these users could be considered technically as intermediate users, charged with translating or disseminating the forecast information to end users. End users of forecasts can be found in a variety of sectors, such as agriculture, water resources, electricity supply, transport, construction, crop and other insurance industries, recreation, nature conservation, and planning.

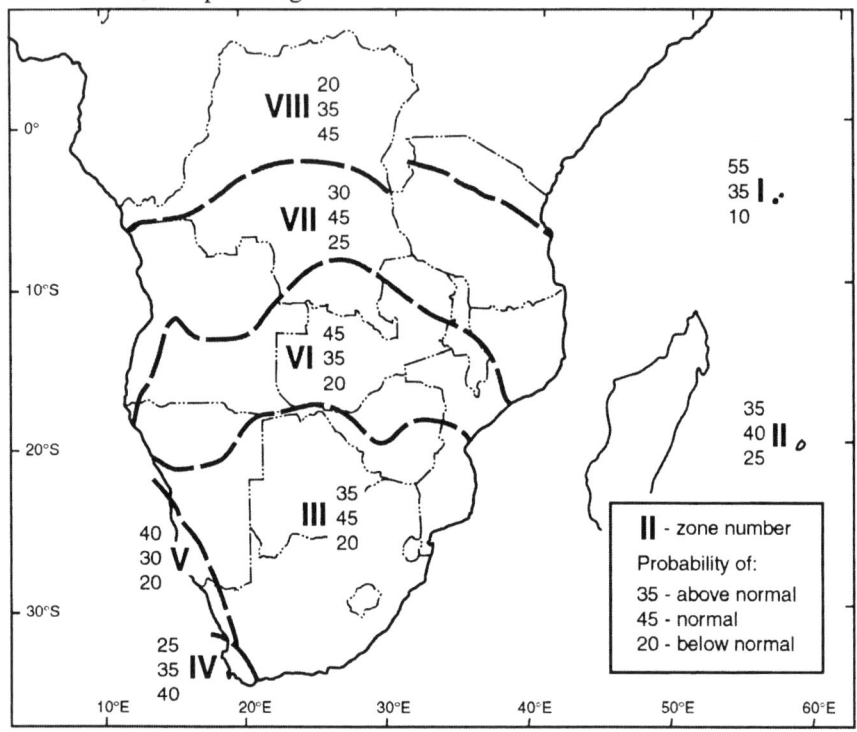

Source: Drought Monitoring Centre, 2002

Figure 1.2   January–March 2000 rainfall forecast for SADC region

To date, assessments of the benefits of seasonal forecasts to end users have been undertaken primarily within the transport, agriculture, water, and energy sectors (GCOS, 1995). Estimating the value of forecasts is a difficult task because of the wide range of user perceptions, uptake, and needs of forecasts. There are two general approaches used to assess the value of forecasts, namely prescriptive and descriptive (GCOS 1995; Stewart, 1997; Stern and Easterling, 1999). Prescriptive approaches usually assume that users of such information behave in a manner that is optimal according to some normative theory of decision making (Stewart, 1997). These approaches use formal models of individual users' decision-making processes for economic sectors influenced by the aggregate effects of individuals' decisions (Stern and Easterling, 1999). Descriptive approaches focus on the actual behavior of users and their actual information-processing and decision-making procedures. Most of the assessments of the use and value of seasonal forecasts carried out in the case studies described in this book are of the descriptive type.

The forums and forecasts have increased public awareness about climate forecasting and its role in decision making. While assessments of the need, value, and uptake of forecasts have been positive in some cases (e.g., Sonka Changnon, and Hofing, 1988; Katz and Murphy 1997; IRI, 2001a), several factors continue to impede forecast uptake and use. Some of these factors include: 1) the information is too broad and non-specific in terms of spatial scale; 2) information available is too general and difficult to interpret; 3) predictions for a given period are issued too late for use; 4) predictions for discrete and critical periods are not available; and 5) lack of access to experts to interpret predictions (Changnon, 1992; Stern and Easterling, 1999; Glantz, 2000; Letson et al., 2001).

Reaching end users in southern Africa and providing them with forecasts that are of value is one of the major challenges to the success of the SARCOF process. Some of these concerns related to forecast use were discussed at the Workshop on User Responses to Seasonal Climate Forecasts in Southern Africa, held in Dar es Salaam, Tanzania in 1999 (O'Brien et al., 2000). Three issues emerged as important: The first relates to the degree and level of "fit" between the users and producers of forecasts. These two communities usually work in isolation from each other, yet are expected through forecasts to find mutually beneficial uses and products. Failure to fully understand end users and their needs and constraints, including the institutional channels through which information flows to them, how information is owned and acted upon, and issues such as credibility and support systems to ensure ownership and use of information, are further limiting factors in forecast uptake and use.

The second issue, closely related to the first, is communication and the development of information channels. It was acknowledged that the channels of dissemination from meteorological services to intermediate and end users remain weak throughout the region. In some countries, agricultural extension systems can promote the dissemination of forecasts, whereas in other countries, they are less effective. Although the radio is the most widely used media for forecast distribution in southern Africa, informal networks also play an important role in information dissemination.

The third issue focuses on the ability of end users to make use of the forecasts. To have value, forecasts must lead to actions that would otherwise have not been taken (Stern and Easterling 1999). In southern Africa, even if perfect forecasts were disseminated in an optimal manner, there remain significant factors constraining their use and thus limiting their value. This situation is not unique, and others working on the dissemination and uptake of forecasts elsewhere in the world have noted similar constraints (Broad and Agrawala, 2001; Eakin, 2000; Nelson and Finan, 2000; Hammer, Nicholls, and Mitchell, 2000; Roncoli, Ingram, and Kirshen, 2000, 2001; Ingram, Roncoli, and Kirshen, 2002). Constraints are in many cases related to the limited options available to farmers, such as alternative seeds, draft power, irrigation, or availability of land. Identifying such constraints is seen as essential in understanding how coping and adaptive strategies can be affected and strengthened by seasonal climate forecasts (O'Brien et al., 2000; Vogel, 2000).

The chapters that follow summarize some of the experiences and problems that researchers have identified in relation to the dissemination and use of seasonal forecasts in southern Africa. Though the focus is limited to southern Africa, similar issues have emerged in other regions of the world where seasonal climate forecasts can serve as a potentially valuable source of information. By synthesizing these case studies on forecast dissemination and use, and extracting useful insights on the end-to-end forecasting process, we hope to identify ways to improve strategies for coping with climate risk in the region.

Overview of Chapters

In Chapter 2, Maxx Dilley examines progress related to the use of seasonal climate forecasts at the regional level. As Science Advisor in the Office of U.S. Foreign Disaster Assistance at the U.S. Agency for International Development (U.S. AID) during the 1990s, Dilley was in a position to directly observe the extent to which climate forecasts influenced decision-making in the disaster relief community. By comparing responses related to three El Niño events in southern Africa (1991/92, 1994/95, and 1997/98), he identifies a transition taking place within the international aid community, whereby emphasis has shifted from monitoring to prediction, and from post-disaster relief to preparedness and prevention.

In 1997/98, international and regional agencies (and in some cases national-level agencies) took advantage of increased knowledge of climate impacts and the availability of seasonal forecasts. Actions to reduce vulnerability, such as shifting crop mixes or including drought-tolerant varieties, were developed by U.S. AID's Famine Early Warning System (FEWS). Donors and other international agencies developed timelines and contingency plans in preparation for a regional food crisis precipitated by drought. While such a drought did not materialize, the experience made valuable contributions to a broader learning process.

Although the 1997/98 climate forecasts led to a more proactive approach to disaster management, Dilley points out that forecasts are only one factor among many influencing livelihoods and sustainable development in southern Africa. He argues that although there is value in continuing to improve preparedness at the international and regional levels, the crux of the effort should focus on communicating currently available information to the grass-roots level. He also warns that prospects for the immediate future have been set by recent events, which in the case of the 1997/98 El Niño led to some misguided advice and in some cases a public backlash against the forecasts.

Chapter 3 focuses on national level responses to seasonal climate forecasts. Focusing on the cases of Namibia, Zambia, and Malawi during the 1997/98 El Niño event, Anne Thomson examines how the public sector responded to the seasonal forecasts. One of the issues she addresses is the scope for potential action available to governments in southern Africa. She notes that this scope has been limited over the past two decades, concurrent with decreasing state control over agricultural production and marketing associated with structural adjustments and market liberalization. Governments are now less likely to import food or impose trade restrictions. Instead, activities are now largely limited to information activities, research, and extension services. Consequently, governments might be expected to play an important role in forecast dissemination, and in providing information on appropriate risk-reducing responses. They might also be expected to undertake contingency planning to ensure adequate food supply under a drought situation.

By examining government actions in 1997, Thomson identifies actions taken, as well as missed opportunities. In terms of forecast dissemination, she found that the governments of Malawi, Namibia, and Zambia overall did a reasonable job in informing organizations and stakeholders in major cities, but a poor job in getting the information out to small-scale farmers. Potential channels for disseminating forecasts to small-scale farmers include farmer unions and extension services, but these channels are constrained by poor funding, limited capacity, and limited contacts.

Not all governments were willing or able to effectively provide advice to farmers, and with the exception of Malawi, few efforts for contingency planning were undertaken. A "wait and see" attitude was prevalent, in part because the forecasts have not yet proven themselves to be reliable, and in part because the probability-based forecasts were found difficult to operationalize. Interestingly, Thomson found that financial constraints were seldom as much of a problem as uncoordinated and underdeveloped strategies for using the forecasts. Thomson identifies a number of potential areas that could contribute to improved forecast responses, including increased networks for information flow.

While Chapters 2 and 3 highlight the new challenges and opportunities associated with pre-season interventions based on climate forecasts, Chapter 4 exposes some of the limitations, particularly for small-scale farmers. Here, Roger Blench argues that despite the interest and enthusiasm for seasonal forecasts, there is a considerable gap between the information that is provided by meteorological services and the information that is useful to small-scale farmers. He stresses that the type of information of interest to farmers, such as the distribution and timing of rainfall, are not included in seasonal forecasts. Furthermore, the probabilistic information that is available applies to a coarse area, and it cannot be easily tested for accuracy. Farmers thus run the risk of using a forecast that is technically correct, but nevertheless encourages inappropriate responses.

Regarding dissemination, Blench notes that forecasts are currently targeted at an oversimplified dichotomy of farmers; namely commercial and subsistence farmers. He instead proposes the development of a more detailed user profile. Such profiles would consider not only the scale of enterprise, but also the type of production strategy, degree of dependence on rainfall, access to safety nets, level of education, and access to information. Blench then argues that agricultural extension services are inappropriate channels for forecast dissemination, and instead suggests that using farmers' existing information networks would be more effective.

Blench is skeptical to the assumption that, given a correct forecast, farmers will adopt a strategy that reflects the predicted climate conditions. He argues that rural producers do not gamble all of their resources on a single strategy, and instead prefer to spread risk by managing diversity. In short, he argues that seasonal forecasts are not suited to the way farmers actually farm in most of Africa. The forecasts will not be useful, he argues, until a more integrated approach for dissemination is developed, and the production strategies and needs of small-scale farmers are explicitly considered.

The second part of the book focuses on specific case studies covering a wide range of user groups in the agricultural sector. In Chapter 5, Jerry Hudson and Coleen Vogel present results from a study on forecast use among livestock farmers in the western parts of the North-West Province of South Africa, where seasonal climate forecasts offer the potential for advanced planning in the livestock sector. For example, farmers can purchase additional fodder in advance, make arrangements for grazing on additional pastureland, or sell animals while market prices are still high, and while animals are in good condition. Nevertheless, among the farmers interviewed, the majority viewed forecasts as being not at all or only somewhat valuable. Commercial farmers were almost twice as likely as communal farmers to hold this view, and less than one-third of the farmers interviewed would make management decisions based on climate forecasts. Of the farmers who heard forecasts for the 1997/98 El Niño and took actions, a small percentage incurred losses because they reduced animal size in anticipation of a drought that did not materialize.

By comparing drought responses among both commercial and communal farmers in this arid region, significant differences in coping strategies were identified. Whereas commercial farmers tend to decrease herd size under drought conditions and maintain the weight of the remaining livestock, communal farmers tend to maintain herd size and decrease the per-animal weight. Hudson and Vogel point out that these strategies emerge as a result of interacting environmental and socio-economic factors that have evolved to minimize anticipated effects of recurrent droughts. These interacting factors are dynamic and complex, and go beyond typical cultural stereotypes, whereby commercial farmers are seen as successful animal producers who make optimal use of available resources, while communal farmers are seen as practicing inefficient animal husbandry (characterized by overgrazing and land degradation), keeping animals instead largely as measures of wealth and social status.

Climate forecasts are currently not viewed as valuable input to livestock production because of their format and perceived lack of precision. Nevertheless, they are still considered potentially useful and thus desirable information for farmers. Differences in coping strategies suggest that climate forecasts can potentially be used to plan for the sale of livestock among commercial farmers, and for the purchase of fodder among communal farmers. This supports calls for improved targeting of forecasts to specific user groups, with special attention to user needs and the timing of forecasts.

In Chapter 6, Louise Bohn examines both the potential and constraints to forecast use among agribusinesses in Swaziland. Focusing on sugar cane estates, a forestry plantation, a citrus-fruit grower, and a cattle and game ranch, Bohn uses timelines to identify the timing of specific activities for each agribusiness, the major decisions that must be made regarding these activities, and the climate information that can be used and is needed. The timelines help to identify where interventions can be made based on climate information, thus they call attention to the potential value of seasonal forecasts in agribusinesses.

While the potential value of climate forecasts for agribusinesses is high, Bohn also identifies a number of constraints that diminish the value of forecasts. The constraints are related to the timing of the forecasts, the spatial scale of the information, the risk and uncertainty, and the type of event that is forecast. Most of these constraints relate to the actual characteristics of forecasts. However, there are some internal constraints related, for example, to budgets and planning. Despite access to forecasts and resources to respond, it is difficult for agribusinesses to flexibly adjust production as a result of probabilistic climate forecasts. If some of the forecast constraints were addressed, it is likely that agribusinesses both could and would make use of seasonal forecasts.

There are clearly opportunities within the commercial agricultural sector to make use of seasonal climate forecasts. However, a large part of southern Africa's farmers operate at a small-scale, farm marginal lands, and lack access to resources such as improved seeds, irrigation, and draft power. The potential for using seasonal climate forecasts may be constrained for these farmers, not only because they do not receive the information, but also because they have limited response options compared to commercial farmers.

In Chapter 7, Jennifer Phillips explores determinants of forecast use among communal farmers in Zimbabwe. She looks specifically at the role that access to resources, or "wealth," plays in determining the use of climate forecasts among smallholders. Phillips conducted two household surveys in different Natural Regions of Zimbabwe during the 1997/98 and 1998/99 ENSO events to gather data on household assets, climate information, local customs for rainfall prediction, and planting strategies for the coming season. For the analysis, she stratified households into three groups based on asset levels. Data were also stratified according to Natural Region, and according to the type of year that was forecast (El Niño versus La Niña).

The results indicate that asset levels do influence access to climate information, particularly when dissemination campaigns are moderate, as in the case of the 1998/99 La Niña event. However, asset levels did not play a significant role in determining the response to the forecast (i.e., choices of crops or planting dates). Instead, responses related to crop management were more closely linked to the underlying agroecological conditions. Farmers in relatively wet regions cut back on the area planted in anticipation of too much rainfall in 1998/99, whereas farmers in dry regions planted significantly more maize than normal to capture the benefits of an expected good rainfall year. The results from Zimbabwe show that climate information related to a good year (i.e., higher than normal rainfall) is useful to farmers, and could potentially yield greater benefits to smallholders than forecasts for below normal conditions. This underscores a need to actively disseminate forecasts in "non-crisis" years.

In Chapter 8, Channing Arndt, Melanie Bacou, and Antonio Cruz address economic perspectives of climate forecasts, using Mozambique as a case study. Identifying the economic value of forecasts is seen as critical for justifying long-term funding for forecast development. Nevertheless, it is challenging to derive dollar values for the forecasts. In fact, although desirable, such values may be misleading, given the complexity of agricultural decision-making and the downstream consequences of climate variability for other sectors. The results presented in Chapter 8 point to the agricultural marketing system as a major potential source of gains from climate forecasts.

The authors note that in order to be valuable, the forecasts must induce economic agents to take actions to improve economic conditions, above and beyond what would have been done in the absence of a forecast. Using a computable general equilibrium (GCE) model, Arndt, Bacou, and Cruz examine linkages and feedback effects between farmer responses, marketing sector responses, and non-agricultural sector responses. With such an approach, the results are driven largely by the structure of the economy, particularly the relative size of productive sectors, the importance of international trade, and the nature of household consumption.

The authors conducted three experiments based on this approach. A comparison of welfare measures shows that rural households disproportionately bear the costs of drought, even if responses to a perfect forecast are taken into consideration. Although small-scale farmers can adjust crop mixes or areas planted in response to forecasts, the scope for response is relatively limited. Moreover, the benefits of these responses accrue disproportionately to urban consumers, through price changes and differential consumption patterns. In contrast, they find that gains to the marketing sector are much more significant. In fact, even small gains in the efficiency of the marketing system in response to climate forecasts could lead to substantial improvements in welfare.

The third part of the book addresses more specific aspects of user needs for climate information and considers the future role of forecasts in coping with climate variability. In Chapter 9, Nganga Kihupi and his coauthors discuss the impacts of the 1997/98 El Niño event on Tanzania. This event led to catastrophic flooding in many parts of Tanzania. The seasonal climate forecasts were heard by relatively few of the farmers interviewed. Although the El Niño phenomenon was unfamiliar to many prior to the widespread media coverage of 1997/98, the authors point out that farmers have always maintained a wide range of traditional indicators used to predict climate conditions for the upcoming season. This indigenous knowledge has been passed down over generations, and is generally more trusted than the new seasonal forecasts disseminated by meteorological services. However, the various indicators have not been quantified or used to develop formal predictions.

The authors contend that many of the indigenous indicators are biophysical responses to atmospheric conditions, such as nighttime temperatures, humidity, wind speed, and so on. In light of the climate information needed by farmers, which includes the onset and cessation of rains, the length of the growing season, and the occurrence of dry spells, the authors developed an Aridity Index that can be correlated to data from local meteorological stations. Significant correlations can be used to identify which meteorological factors are the best predictors of upcoming climate conditions. Indigenous indicators can be considered proxies for these factors. The authors suggest this method as one way to integrate local indicators and indigenous knowledge with seasonal climate forecasts derived from atmosphere-ocean models.

Commercial user needs are examined in Chapter 10, based on a survey of the value of seasonal climate forecasts in South Africa carried out by Emsie Klopper and Anna Bartman. As employees of the South African Weather Service, the authors are interested in optimizing the use and value of seasonal forecasts. Indeed, from the perspective of the South African Weather Service, which was privatized in 2001, seasonal climate forecasts represent a product with an existing demand and a strong potential for future growth. The authors conducted the surveys from 1997 to 2000 to identify both user requirements and constraints to forecast use. The survey targeted commercial farmers, as they represent the largest client group of the Weather Service. Additionally, they generally have access to information and infrastructure to make maximum use of the forecasts. The results of the survey show that commercial farmers have indeed made use of the forecasts in production-related decisions, but that forecasts would be considered more useful if they were targeted to smaller geographical regions. The results also demonstrate that a wider array of forecasts of different climate indicators is desired. In order to develop more tailored products, greater communication between forecast providers and user groups is necessary.

The mission of the Weather Service is to "provide an internationally respected and efficient service to help safeguard life and property and improve socio-economic structures for the benefit of all South Africans and with a particular focus on clients and disadvantaged communities" (SAWS, 2002). Toward this goal, there is a need to target climate forecasts at different user groups, to tailor climate forecasts to particular user groups needs, and to package the forecasts so that they can be easily understood. Yet the authors stress that climate information is just one factor among many that influence farmers' decisions, and that the value of climate forecasts must be assessed within a larger context that includes economic factors, market conditions, and the socio-political situation of the country.

One theme that recurs in different chapters is the mismatch between the information provided by seasonal climate forecasts and the needs of various user groups. In Chapter 11, Neil Ward and Jolamu Nkhokwe confront this mismatch by considering the feasibility of generating the types of information requested by users in Malawi. Both authors are climate scientists active in the production of seasonal forecasts. Through participation in the long-lead forecast component of the Malawi Environmental Management Project, the authors drew upon a survey of user needs and interviewed potential users to discuss their individual requirements. Representatives from an NGO concerned with food security, an agricultural association, a regional water board, and a tour operator provided insights to the types of information that could be included in "targeted packages." More than a general seasonal climate outlook, these users need downscaled forecasts and information about the evolution of the season, as well as monitoring of current conditions. They could also benefit from historical climate information and additional variables, such as sunshine hours or water budgets.

Ward and Nkhokwe then present some graphics that illustrate how seasonal forecasts can be downscaled to specific locations, and how they can provide information on the probable evolution of the season. They show that targeted packages of climate information are indeed feasible. However, they argue that it is up to the meteorological services to identify the balance between development of general products such as seasonal forecasts and targeted products aimed at diverse user groups.

In Chapter 12, we conclude by considering the future of seasonal climate forecasts. Although forecasts have been heralded as a promising tool for coping with climate stresses and shocks, there is a considerable gap between the potential value of forecasts and their actual value in southern Africa. This gap can be attributed in part to issues related to dissemination, access, and relevance of forecast information. Equally significant, however, is the limited ability of many potential users to take actions in response to forecast information, because of limited resources, lack of access to agricultural inputs and credit, and other constraints. We argue that climate forecasts must be discussed in the context of coping strategies and adaptation to climate variability and long-term change. In particular, there is a need for widening the discourse on seasonal climate forecasts to include the political and economic factors that shape and enable coping and adaptation.

We point out that economic changes taking place at the global and national scale are influencing the ability of many southern Africans to cope with climate variability. This, along with the spread of HIV/AIDS, challenges to institutional capacity, conflicts and so forth, creates a dynamic context for forecast use. Understanding this context, we argue, is a prerequisite for making optimal use of seasonal forecasts. Research on the production and uptake of forecasts needs to be coupled to increased efforts to understand how users currently manage their systems and cope with a variety of stresses and shocks.

The ultimate objective of the seasonal forecast process is to meet end-user needs. To date, only a limited representation of potential users in southern Africa have participated in the forecast process, or have heard the results of this process. There remains a wide range of unheard voices and unaddressed needs when it comes to climate information. As the science of climate forecasting moves forward, and as the understanding of user needs improves, it is critical to develop a coherent strategy for expanding the use and increasing the benefits of seasonal climate forecasts.

Notes

1   Two types of forecasts can be distinguished: those issued for short-term periods ranging from one to several days (i.e., weather forecasts) and those that provide an outlook of the climate system over several months (i.e., seasonal climate forecasts).

2   The *Workshop on User Responses to Seasonal Climate Forecasts in Southern Africa* was held in Dar Es Salaam, Tanzania from September 10-11, 1999 (see O'Brien et al., 2000). The objective of the workshop, was to present, discuss and compare research, primarily in relation to the agricultural sector of southern Africa.

3   SADC was established at the Summit of Heads of State or Government on July 17, 1992, in Windhoek, Namibia. It developed from the SADCC (Southern African Development Co-ordination Conference), which was established by nine southern African states in 1979 to pursue policies aimed at economic liberation and integrated development of national economies (SADC, 2002).

4   In the southern Africa region, several groups release forecasts and have expertise in providing forecasts. These groups include the Drought Monitoring Centre (DMC) in Harare, which monitors global forecasts as they emerge and issues modified summary forecasts before the rainy season (Gibberd et al., 1995), as well as a few groups in South Africa, such as the University of Zululand, the University of Cape Town and the South African Weather Service.

5   The Southern African Regional Climate Outlook Forum (SARCOF) was an outcome of the *Workshop on Reducing Climate-Related Vulnerability in Southern Africa*, held October 1-4, 1996 in Victoria Falls, Zimbabwe.

6 Although the process initially included mid-season meetings to correct and update forecasts and post-season meetings to evaluate forecasts, in recent years SARCOF has been limited to one annual meeting.

References

Allan, R.J., Lindesay, J.A. and Parker, D.E. (1996), *El Niño Southern Oscillation and Climatic Variability*, CSIRO Publishing, Melbourne.

Archer, E. (2001), 'Forecast Use, Agricultural Production and Gender: Mangondi Village, Northern Province, South Africa', Paper presented at IRI workshop on communication of climate forecast information, 6-8 June 2001, Palisades, NY.

Barnston, A.G., van den Dool, H.M., Zebiak, S.E., Barnett, T.P., Ji, M., Rodenhuis, D.R., Cane, M.A., Leetmaa, A., Graham, N.E., Ropelewski, C.R., Kousky, V.E., O'Lenic, E.A., and Livezey, R.E., (1994), 'Long-lead Seasonal Forecasts – Where do we Stand?', *Bulletin of the American Meteorological Society*, vol. 75, pp. 2097-2114.

Battisti, D.S. and Sarachik, E.S. (1995), 'Understanding and Predicting ENSO', *Review of Geophysics Supplement*, vol. 33, pp. 1367-1376.

Benson, C. and Clay, E., (1998) *The Impact of Drought on Sub-Saharan Economies: A Preliminary Examination*, World Bank Technical paper No. 401, World Bank, Washington, DC, USA.

Broad, K. and Agrawala, S. (2001), 'The Ethiopia Food Crisis: Uses and Limits of Climate Forecasts, *Science,* vol. 289, pp. 1693-1694.

Buizer, J., Foster, J. and Lund, D. (2000), 'Global Impacts and Regional Actions: Preparing for the 1997-98 El Niño', *Bulletin of the American Meteorological Society*, vol. 81, no. 9, pp. 2121-2139.

Cane, M.A. (2000), 'Understanding and Predicting the World's Climate System', in G.L. Hammer, G.L., N. Nicholls and C. Mitchell (eds), *Applications of Seasonal Climate Forecasting in Agricultural and Natural Ecosystems, The Australian Experience*, Kluwer Academic Publishers, Dordrecht, pp. 29-50.

Cane, M.A., Eshel, G. and Buckland, R.W. (1994), 'Forecasting Zimbabwean Maize Yield using Eastern Pacific Sea Surface Temperatures', *Nature*, vol. 370, pp. 204-205.

Cane, M.A., Zebiak, S.E. and Dolan, S.C. (1986), 'Experimental Forecasts of El Niño', *Nature*, vol. 321, pp. 827-832.

Changnon, S.A. (1992), 'Contents of Climate Predictions Desired by Agricultural Decision Makers', *Journal of Applied Meteorology*, vol. 76, pp. 711-720.

Climate Prediction Center (1992–7), *Experimental Long-lead Forecast Bulletin*, Climate Prediction Center, W/NP5, NOAA Science Center, Washington D.C.

Dilley, M. (2000), Reducing Vulnerability to Climate Variability in Southern Africa: The Growing Role of Climate Information', *Climatic Change*, vol. 45, pp. 63-73.

Dorrington, R., Bourne, D., Bradshaw, D., Laubscher, R. and Timaeus, I.M. (2001), *The Impact of HIV/AIDS on Adult Mortality in South Africa*, Technical Report, Burden of Disease Research Unit, South African Medical Research Council, South Africa.

Drought Monitoring Centre (2002), Southern Africa Regional Climate Outlook Forum (SARCOF), http://www.dmc.co.zw/.

Eakin, H. (2000), 'Smallholder Maize Production and Climatic Risk: a Case Study from Mexico', *Climatic Change*, vol. 45, pp. 19-36.

Garanganga, B.J. (1998), 'Review of Southern Africa Climate Variability,' in M. Harrison (ed.), *First Report of the Enrich Southern Africa Regional Climate Outlook Forum to the European Commission*, available at http://151.170.240.7/sec5/NWP_seasonal/ report1 /index.html.

Gibberd, V., Rook, J., Sear, C.B., and Williams, J.B. (1995), *Drought Risk Management in Southern Africa: the Potential of Long Lead Climate Forecasts for Improved Drought Management*, Natural Resources Institute, prepared for Overseas Development Administration and the World Bank, Kent.

Glantz, M.H. (1996), *Currents of Change: El Niño's Impact on Climate and Society*, Cambridge University Press, Cambridge.

Glantz, M.H. (2000), *Once Burned, Twice Shy? Lessons Learned from the 1997-98 El Niño*: UNEP/NCAR/UNU/WMO/ISDR Assessment, October 2000.

Glantz, M., Betsill, M. and Crandall, K. (1997), *Food Security in Southern Africa: Assessing the Use and Value of ENSO Information*, NOAA Project, National Center for Atmospheric Research, Environmental and Societal Impacts Group, Boulder, Colorado.

Global Climate Observing System (GCOS) (1995), *The Socio-economic Benefits of Climate Forecasts: Literature review and Recommendations*, prepared by the GCOS Working Group on Socioeconomic Benefits, April, 1995, GCOS 12, WMO/TD No. 674, United Nations Programme and International Council of Scientific Unions.

Goddard, L., Mason, S.J., Zebiak, S.E., Ropelewski, C.P., Basher, R., and Cane, M.A. (2001), 'Current Approaches to Seasonal-to-Interannual Climate Predictions', *International Journal of Climatology*, vol. 21, pp. 1111-1152.

Greischar, L. and Hastenrath, S. (1997), Neural Network to Produce 'Short Rains' at the Coast of East Africa for Boreal Autumn 1997, *NOAA Experimental Long Lead Bulletin*.

Halpert, M.S. and Ropelewski, C.F. (1992), 'Surface Temperature Patterns Associated with the Southern Oscillation', *Journal of Climate*, vol. 5, pp. 577-593.

Hammer, G.L., Nicholls, N., and Mitchell, C. (eds) 2000. *Applications of Seasonal Climate Forecasting in Agricultural and Natural Ecosystems: The Australian Experience*, Kluwer, The Netherlands.

Harrison, M. (1984), 'A Generalized Classification of South African Summer Rain-bearing Synoptic Systems, *Journal of Climatology*, vol. 4, pp. 547-560.

Harrison, M. (1986), *A Synoptic Climatology of South African Rainfall Variations*, Ph.D. Thesis, University of the Witwatersrand, Johannesburg.

Hastenrath, S. (1985), *Climate and Circulation of the Tropics*, Reidel, Dordrecht.

Hulme, M. (ed.) (1990), *Seasonal Rainfall Forecasting for Africa: Recommendations for Forecast Development, Implementation, and Impact Assessment*, Report prepared for the Natural Resources Institute, Kent, Climate Research Unit, Norwich, England.

Hulme, M. (ed.) (1996), *Climate Change and Southern Africa*, Report to WWF International by the Climatic Research Unit, UEA, Norwich.

Ingram, K.C., Roncoli, C., and Kirshen, P. (2002), 'Opportunities and Constraints for Farmers of West Africa to Use Seasonal Precipitation Forecasts with Burkina Faso as a Case Study', *Agricultural Systems,* in press.

International Fund for Agricultural Development (IFAD) (2001), *Rural Poverty Report 2001—The Challenge of Ending Rural Poverty*, International Fund for Agricultural Development, Rome.

International Research Institute for Climate Prediction (IRI) (2001a), *Coping with the Climate: A Way Forward, Preparatory Report and Full Workshop Report*, a multi-stakeholder review of Regional Climate Outlook Forums concluded at an international workshop, 16-20 October 2000, Pretoria, South Africa.

International Research Institute for Climate Prediction (IRI) (2001b), *Coping with the Climate: A Way Forward, Summary and Proposals for Action*, a multi-stakeholder review of Regional Climate Outlook Forums concluded at an international workshop, 16-20 October 2000, Pretoria.

Janowiak, J.E. (1988), 'An Investigation of Interannual Rainfall Variability in Africa', *Journal of Climate*, vol. 1, pp. 240-255.

Jury, M.R. (1995), 'A Review of Research on Ocean-atmosphere Interactions and South African Climate Variability', *South African Journal of Science*, vol. 91, pp. 289-294.

Katz, R. and Murphy, A. (1997), *Economic Value of Weather and Climate Forecasts*, Cambridge University Press, New York.

Kirtman, B.P. and Schopf, P.S. (1998), 'Decadal Variability in ENSO Predictability and Prediction', *Journal of Climate*, vol. 11, pp. 2804-2822.

Klopper, E. (1999), The Use of Seasonal Forecasts in South Africa During the 1997/98 Rainfall Season, *Water SA*, vol. 25, pp. 311-316.

Landman, W.A. and Mason, S.J. (1999), 'Change in the Association between Indian Ocean Sea-surface Temperatures and Summer Rainfall over South Africa and Namibia', *International Journal of Climatology*, vol. 19, pp. 1477-1492.

Landman, W.A. and Mason, S.J. (2001), 'Forecasts of Near-global Sea Surface Temperatures using Canonical Correlation Analysis', *Journal of Climate*, vol. 14, pp. 3819-3833.

Latif, M., Anderson, D.L.T., Barnett, T.P., Cane, M.A., Kleeman, R., Leetmaa, A., O'Brien, J., Rosati, A. and Schneider, E. (1998), 'A Review of the Predictability and Prediction of ENSO', *Journal of Geophysical Research*, vol. 103, pp. 14375-14393.

Leichenko, R.M. and O'Brien, K.L. (2002), 'The Dynamics of Rural Vulnerability to Global Change: The Case of Southern Africa', *Mitigation and Adaptation Strategies for Global Change*, vol. 7, pp. 1-18.

Letson, D., Llovet, I., Podesta, G., Royce, F., Brescia, V., Lema, D., and Parellada, G. (2001), User Perspectives of Climate Forecasts: Crop Producers in Pergaminao, Argentina', *Climate Research*, vol. 19, pp. 57-67.

Lindesay, J.A. (1988), 'Southern African Rainfall, the Southern Oscillation and a Southern Hemisphere Semi-annual Cycle', *Journal of Climatology*, vol. 8, pp. 17-30.

Lindesay, J.A. (1998), 'Present Climates of Southern Africa', in J.E. Hobbs, J.A. Lindesay, and H.A. Bridgman, (eds), *Climates of the Southern Continents, Present, Past and Future*, Wiley, New York, pp. 5-62.

Lindesay, J.A., Harrison, M.S.J., and Haffner, M.P. (1986), 'The Southern Oscillation and South African Rainfall', *South African Journal of Science*, vol. 82, pp. 196-189.

Makarau, A. and Jury, M. (1997), 'Predictability of Zimbabwe summer rainfall', *International Journal of Climatology*, vol.17, no. 1, pp. 1421-1432.

Mason, S. (1997), 'Review of recent developments in seasonal forecasting of rainfall', *Water SA*, vol. 23, pp. 57-62.

Mason, S.J., Goddard, L., Graham, N.E., Yulaeva, E., Sun, L. and Arkin, P. (2000), 'The IRI Seasonal Climate Prediction System and the 1997/98 El Niño Event', *Bulletin of the American Meteorological Society*, vol. 80, pp. 1853-1873.

Mason S.J. and Goddard L. (2001), 'Probabilistic Precipitation Anomalies Associated with ENSO', *Bulletin of the American Meteorological Society*, vol. 82, pp. 619-638.

Mason, S.J. and Jury, M.R. (1997), 'Climatic Change and Inter-annual Variability over Southern Africa: A Reflection on Underlying Processes', *Progress in Physical Geography*, vol. 21, pp. 23-50.

Mason S.J. and Mimmack, G.M. (2002), 'Comparison of Some Statistical Methods of Probabilistic Forecasting of ENSO, *Journal of Climate*, vol. 15, pp. 8-29.

Matarira, C.H. (1990), 'Drought over Zimbabwe in a Regional and Global Context', *International Journal of Climatology*, vol. 10, pp. 609-625.

McCarthy, J. Canziani, O.F., Leary, N.A., Dokken, D.J. and White, K.S. (2001), *Climate Change 2001: Impacts, Adaptation, and Vulnerability*. Contribution of Working Group II to the Third Assessment Report of the Intergovernmental Panel on Climate Change, Cambridge University Press, Cambridge, England.

McPhaden, M.J., Busalacchi, A.J.,Cheney, R., Donguy, J. R., Gage, K.S., Halpern, D., Ji, M., Julian, P., Meyers, G., Mitchum, G.T., Niiler, P.P., Picaut, J., Reynolds, R.W., Smith, N., and Takeuchi, K. (1998), 'The Tropical Ocean - Global Atmosphere Observing System: A Decade of Progress', *Journal of Geophysical Research*, vol. 103, pp. 14169-14240.

National Oceanic and Atmospheric Administration, Office of Global Programs (NOAA-OGP) (1999), An Experiment in the Application of Climate Forecasts: NOAA-OGP Activities Related to the 1997-98 El Niño Event. Office of Global Programs, NOAA, Dept of Commerce, a publication of the University Corporation for Atmospheric Research, Washington, D.C.

Neelin, J.D., Battisti, D.S., Hirst, A.C., Jin, F.F., Wakata, Y., Yamagata, T., and Zebiak, S.E. (1998), 'ENSO Theory', *Journal of Geophysical Research*, vol. 103, pp. 14261-14290.

Nelson, D.R. and Finan, T.J. (2000), 'The Emergence of a Climate Anthropology in Northeast Brazil', *Practicing Anthropology*, vol. 22, pp. 6-10.

O'Brien, K., Sygna, L., Naess, R., Kingamkono, R. and Hochobeb, B. (2000), *Is information Enough? User Responses to Seasonal Climate Forecasts in Southern Africa*. CICERO Report 2000:3, Oslo, Norway.

Orlove, B.S. and Tosteson, J.L. (1999), 'The Application of Seasonal to Internannual Climate Forecasts Based on El Niño-Southern Oscillation (ENSO) Events: Lessons from Australia, Brazil, Ethiopia, Peru and Zimbabwe', Berkeley Workshop on Environmental Politics Working Paper, No. WP, 99-3, Institute of International Studies, University of California, Berkeley, California.

Philander, S.G. (1983), 'El Niño Southern Oscillation Phenomenon', *Nature*, vol. 302, pp. 295-301.

Philander, S.G. (1990), *El Niño, La Niña and the Southern Oscillation*, Academic Press, San Diego.

Phillips, J., Makaudza, E. and Unganai, L. (2001), 'Current and Potential Use of Climate Forecasts for Resource-poor Farmers in Zimbabwe', *Impacts of El Niño and Climate Variability on Agriculture*, American Society of Agronomy Speical Publication Series, vol. 63, pp. 87-100.

Rasmusson, E.M. and Carpenter, T.H. (1982), 'Variations in Tropical Sea Surface Temperature and Surface Wind Fields Associated with the Southern Oscillation/El Niño', *Monthly Weather Review*, vol. 11, pp. 354-384.

Roncoli, C, Ingram, K. and Kirshen, P. (2000), 'Can Farmers of Burkina Faso use Rainfall Forecasts?', *Practicing Anthropology*, vol. 22, pp. 24-28.

Roncoli, C., Ingram, K. and Kirshen, P. (2001), 'The Costs and Risks of Coping with Drought: Livelihood Impacts and Farmer's Responses in Burkina Faso', *Climate Research*, vol. 19, pp. 119-132.

Ropelewski, C.F. and Halpert, M.S. (1987), 'Global and Regional Scale Precipitation Patterns Associated with the El Niño/Southern Oscillation (ENSO)', *Monthly Weather Review*, vol. 115, pp. 1606-1626.

Ropelewski, C.F. and Folland, C.K. (2000), 'Prospects for the Prediction of Meteorological Drought', in D. Wilhite (ed.), *Drought Volume 1, A Global Assessment*, Routledge Hazards and Disasters Series, New York, pp. 21-40.

Sharma, M., Burton, I., van Aalst, M., Dilley, M. and Acharya, G. (2001), *Reducing Vulnerability to Environmental Variability*, Background Paper for the Bank's Environmental Strategy, World Bank, Washington D.C.

Sonka, S.T., Changnon, S.A. and Hofing, S. (1988), 'Assessing Climate Information Use in Agribusiness II: Decision Experiments to Estimate Economic Value', *Journal of Climate*, vol 1, pp. 766-774.

South African Weather Service (SAWS) (2002), 'Vision and Mission', http://www.weathersa.co.za/sawb/mission.html.

Southern African Development Community (SADC) (2002), *About SADC: Background*. http://www.sadc.int/

Stern, P.C. and Easterling, W.E. (1999), *Making Climate Forecasts Matter*, Panel on the Human Dimensions of Seasonal-to-Interannual Climate Variability, Committee on the Human Dimensions of Global Change, National Research Council, National Academy Press, Washington, D.C.

Stewart, T.R. (1997), 'Forecast Value: Descriptive Decision Studies', in R.W. Katz and A.H. Murphy (eds), *Economic Value of Weather and Climate Forecasts*. Cambridge University Press, Cambridge. pp. 147-181.

Stockdale, T.N., Busalacchi, A.J., Harrison, D.E., and Seager, R. (1998), 'Ocean Modeling for ENSO', *Journal of Geophysical Research*, vol. 103, pp. 14325-14355.

Tennant, W.J. (1996), 'Influence of Indian Ocean sea-surface temperature anomalies on the general circulation of southern Africa', *South African Journal of Science*, 92, 289-295.

Thomson, A., Jenden, P. and Clay, E. (1998), *Information, Risk, and Disaster Preparedness: Responses to the 1997 El Niño Event*, Research Report, DFID, SOS SAHEL, London.

Topouzis, D. and du Guerny, J. (1999), Sustainable Agricultural/Rural Development and Vulnerability to the AIDS Epidemic. Joint FAO/UNAIDS publication, Geneva, UNAIDS.

Tyson, P.D. and Preston-Whyte, R. A. (2000), *The Weather and Climate of Southern Africa*, Oxford University Press, Cape Town.

United Nations Development Programme (2000), *Human Development Report 2001: Making New Technologies work for Human Development'*, Oxford University Press, New York.

Vogel, C.H. (2000), 'Usable Science: An Assessment of Long-term Seasonal Forecasts Amongst Farmers in Rural Areas of South Africa', *The South African Geographical Journal*, vol. 82, pp. 107-116.

Walker, S., Mukhala, E., van den Berg, W.J. and Manley, C.R. (2001), *Assessment of Communication and Use of Climate Outlooks and Development of Scenarios to Promote Food Security in the Free State Province of South Africa*, Final report submitted to the Drought Monitoring Centre (Harare, Zimbabwe, University of the Free State and Enviro Vision).

Washington, R. and Downing, T.E. (1999), 'Seasonal Forecasting of African Rainfall: Prediction, Responses and Household Food Security', *The Geographical Journal*, vol. 165, pp. 255-274.

Whiteside, M. (1998), *Living Farms: Encouraging Sustainable Smallholders in Southern Africa*. Earthscan Publications, Ltd., London.

World Bank. (2000), *Can Africa Claim the 21st Century?* The World Bank, Washington, DC.

Zebiak, S.E. and Cane, M.A. (1987), 'A Model El Niño/Southern Oscillation', *Monthly Weather Review*, vol. 115, pp. 2262-2278.

Ziervogel, G. (2001), 'The Role of Seasonal Climate Forecasting in Promoting Sustainable Livelihoods in Southern Africa', Paper presented at the Annual Meeting of the Association of American Geographers, 27 February-3 March 2001, New York.

2 Regional Responses to Climate Variability in Southern Africa[*]

MAXX DILLEY

Introduction

Southern Africa is vulnerable to interannual climate variability. Historically, countries in the region have periodically appealed for international disaster assistance to offset climatic impacts on water supplies, food production and human health. Over 100 disastrous droughts, floods, and related epidemics and pest infestations over the past 30 years have affected approximately 70 million people in the region. (USAID/OFDA, 1996). Climate variability also affects economic performance in sub-Saharan African countries dependent on primary production (Benson and Clay, 1998). Better adaptation to the natural variability in the climate system would reduce risks from climatic hazards and improve use of climate resources.

Southern Africa has considerable potential to make use of emerging seasonal-to-interannual climate forecasting techniques. During El Niño/Southern Oscillation (ENSO) warm events, much of the region tends to be anomalously warm and dry (Ropelewski and Halpert, 1987; Halpert and Ropelewski, 1992). Drought disasters in the region tend to occur in the year following the onset of El Niño and are less frequent at other times (Dilley and Heyman, 1995). Thus knowledge of climate impacts associated with ENSO and monitoring of Pacific sea-surface temperatures provides information for managing drought risks. Furthermore, seasonal climate outlook guidance for southern Africa is now being provided regularly by

[*] This paper is adapted from M. Dilley, (2000), 'Reducing Vulnerability to Climate Variability in Southern Africa: The Growing Role of Climate Information', *Climatic Change*, vol. 45, pp. 63-73. Reprinted with kind permission of Kluwer Academic Publishers.

the Drought Monitoring Center in Harare, supported by national meteorological services, the International Research Institute for climate prediction and other collaborating institutions. Efforts are also on-going to improve the accuracy and reliability of these regional climate forecasts.

In southern African countries dependent on primary production, liberalizing grain markets and trade and implementing other development strategies can help reduce risk from climatic hazards in the long term. Additional short-term adjustments can also be made based on advance information on the likely quality of the region's single October-March agricultural season. In Zimbabwe, where agricultural yields are strongly correlated with Pacific sea-surface temperatures (Cane, Eshel, and Buckland 1994), there are opportunities to increase production under anticipated favorable conditions or reduce losses in dry years through appropriate crop selection and planting practices. Using Zimbabwe as a case study, experimental techniques for modeling potential ENSO impacts on food security have been developed to assist in food security planning (Boudreau, 1997). In South Africa, temperature and rainfall conditions can be used to predict malaria incidence (Jury, 1996), which generates information useful for malaria control efforts (le Seur and Sharp, 1996).

An active learning process is taking place today throughout Africa, characterized by a perceptible shift away from reliance on drought and disaster relief towards better preparedness and prevention (Glantz, 2001). This shift is most evident among international agencies and at the regional level, but extends in some cases to the national level as well. Improved forecasts and better dissemination are some of the ways of reducing risks from extreme events such as floods and droughts.

In this chapter, three El Niño events, occurring in 1991/92, 1994/95 and 1997/98, are compared, with particular reference to the use and uptake of forecast information in advance of the event. All three episodes have been well scrutinized, and a variety of published and unpublished evaluations and assessments are available (see Callihan, Eriksen, and Herrick, 1994; Halmrast-Sanchez et al., 1995; Glantz, Betsill, and Crandall, 1997; Thompson, Jenden, and Clay, 1998; NOAA, 1999; Glantz, 2001). Such a comparison highlights a move among the international relief community beyond monitoring and post-disaster relief, and towards the use of climate forecasts to increase preparedness and prevention strategies. Nevertheless, the analysis also points to a need to enhance risk management strategies at the grass-roots level by communicating the forecast information more widely to the general public.

ENSO Events during the 1990s

Anomalously warm conditions prevailed during the 1990s in the central and eastern tropical Pacific Ocean. The El Niño that developed in 1991 lasted nearly continuously through 1995 and reappeared at record-breaking levels in 1997, following a moderate La Niña in 1995/96. Climate conditions in southern Africa over this period contributed to a series of disasters and disaster responses.

Throughout this period, awareness expanded regarding the impacts of climate fluctuations in southern Africa and the potential for dry conditions during El Niño events. In 1991/92, drought early warning is credited with early implementation of relief programs and avoidance of loss of life (USAID/OFDA, 1992a). Nonetheless, these programs were initiated only after the drought and its impacts were apparent. By 1997, however, El Niño and predicted probabilities of below-normal rainfall were receiving considerable public and government attention in the region prior to the start of the agricultural season.

1991/92

The 1991/92 southern Africa drought led to a major regional disaster. In addition to emergency appeals by national governments, U.S. Ambassadors to eight countries – Lesotho, Malawi, Mozambique, Namibia, South Africa, Swaziland, Zambia and Zimbabwe – determined that the magnitude of the drought had exceeded the countries' ability to cope with it and that international assistance was needed (USAID/OFDA, 1992b). In addition to a massive food-aid operation of over 5 million metric tons of relief food, the region, which is normally a net food exporter, was required to spend an estimated $4 billion on food imports and transport (Callihan, Eriksen, and Herrick, 1994).

Large-scale loss of life was averted during the 1991/92 drought through the resourcefulness and cohesiveness of the affected population (Thompson, 1993) and the timely provision of disaster relief. Relief planning was assisted by early warning systems which monitored the failure of the rains and the subsequent impacts on agricultural production and food supplies. While it was known that the drought was associated with an El Niño event (see Callihan, Eriksen, and Herrick, 1994), and there were theoretically other actions that could have been taken that would have saved money and reduced the threat to food security, knowledge of El Niño did not play a substantial role in mitigating drought impacts. This was in part due to the lack of an organized mechanism for widely distributing the information, and to the fact that potential uses for and users of the information had not been identified in advance (Glantz, Betsill, and Crandall, 1997).

1994/95

Impacts of the 1991/92 drought, the persistence of El Niño, and continued erratic rainfall in southern Africa slowed recovery and contributed to continuing high levels of vulnerability in subsequent years. In 1994/95, a minor El Niño and below-normal rainfall in some areas led to a regional declaration of drought by southern African countries and disaster declarations by U.S. Ambassadors to Lesotho, Swaziland and Zimbabwe. Again, little preventive use was made of climate information and, consequently, the cycle of inadequate rainfall, agricultural production shortfalls, food shortages and emergency relief reoccurred (Dilley, 1996). However, the 1994/95 drought was of smaller magnitude than in 1991/92, and grain market liberalization permitted a more flexible private sector response to meeting grain import and distribution requirements, reducing the need for international emergency assistance.

By this time, there was a growing awareness of the recurring cycle of drought, disaster, relief and recovery (Halmrast-Sanchez et al., 1995). An analysis of events and activities during the 1994/95 El Niño and La Niña of the following year shows the lag between available information and the initiation of response activities (Fig. 2.1). In the 1994/95 season, early warnings gave way to national drought appeals and disaster declarations as it became clear that the situation had exceeded the population's coping ability. As drought relief arrived in late 1995, above-normal rainfall associated with the 1995/96 La Niña triggered disasters caused by storms and flooding.

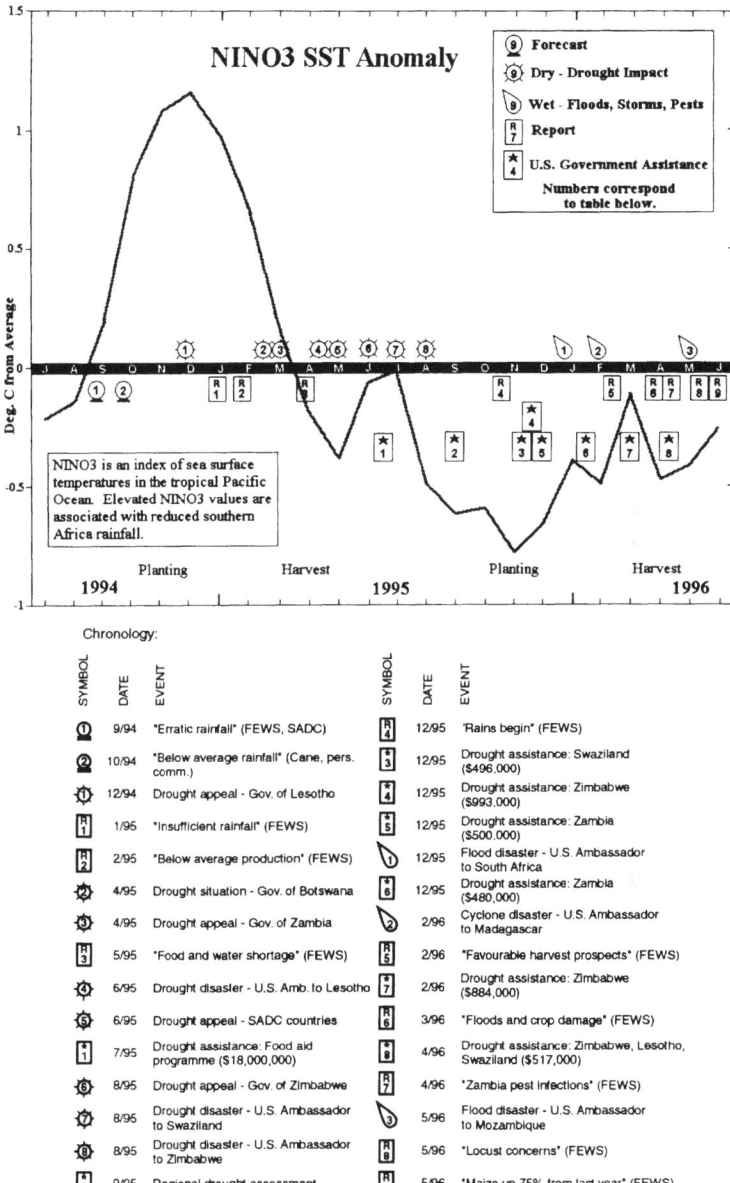

Figure 2.1   Sequence of climate impacts and disaster response in southern Africa from 1994–1996 (based on Dilley, 1996)

Efforts began to systematically develop a process of producing, disseminating, applying and refining seasonal climate forecasts (see NOAA et al., 1996). Research was conducted into the uses for and users of climate forecast information, the level of demand for such information, and the degree of confidence needed to provide a basis for decision-making from the farm- to the national-level (Rook, 1996). During 1995 and 1996, a series of planning meetings laid the groundwork for the production and distribution of consensus climate outlook guidance for southern Africa (see Chapter 1). By the onset of a major El Niño in May 1997, plans were already in place to produce consensus outlook guidance for the 1997/98 season. A forecast forum of national and international climate experts was scheduled for September in Kadoma, Zimbabwe, with a mid-season update planned for December, followed by a post-season evaluation in May 1998.

1997/98

The groundwork of establishing regional forums for consensus climate forecasts (NOAA, 1996) and the sheer magnitude of the 1997/98 El Niño contributed to an unprecedented level of public interest in the phenomenon and its potential impacts months prior to the onset of the 1997/98 agricultural season. A similarly unprecedented public information and planning effort was undertaken in the region to reduce vulnerability to a potential drought. Public and internal government meetings on drought risks and appropriate measures occurred as early as August 1997. National meteorological services participated in the Kadoma climate outlook forum, and some, such as the Zambian Meteorological Service, actively raised public awareness through the media. The Southern Africa Development Community (SADC) Regional Early Warning Unit released bulletins warning of high probabilities of low rainfall. By the time the outlook for the main rains in the region was assessed in Kadoma as probably below-normal, El Niño and the potential for drought had become daily fare in local news media. Following on from the outlook forum of Kadoma, recommendations for enhanced preparation for ENSO-related climate variations were made both internationally and through various regional outlook forums. These recommendations for reducing drought vulnerability were developed by the U.S. Agency for International Development's Famine Early Warning System (FEWS) and included information on shifting crop mixes to include more drought-tolerant varieties and reconsideration of purchases of new animals. In November 1997, still prior to the main rains, SADC convened a high-level policy meeting on the issue of drought and drought management in the region for SADC and donor country representatives.

Outside the region, donors and United Nations agencies developed timelines and contingency plans (see WFP, 1997). Actions included pre-positioning of relief commodities in U.S. ports, establishment of coordination mechanisms, regular monitoring and planning meetings, and limited funding of coordination and information-sharing activities in the region. USAID Missions and FEWS regularly and systematically reported on climate conditions, impacts, and national capabilities to manage a potential emergency.

A widespread, regional drought did not materialize, however, as unexpected Indian Ocean temperature patterns and cyclonic activity overrode ENSO's typical regional effects (Landman and Mason, 1999). For the first part of the season, beginning in October 1997, rainfall was normal to above-normal for most of the region, with below-normal rainfall encountered only in the central part of the region and in central South Africa. From January through March, rainfall tended to be normal over most of the region, with only Namibia experiencing widespread below-normal rainfall (Ward, Unganai, and Garanganga, 1998). Regional agricultural production was 97% of the 1990s average, although in Namibia it was only 61% owing to poor rains (FAO, 1999). In fact, heavy rainfall in the northeast sub-region resulted in a cholera epidemic that led to a declaration of disaster by the U.S. Ambassador to Mozambique. Heavy rains also negatively affected crop production in parts of Zambia. It was the Greater Horn of Africa region to the north that ultimately experienced the worst effects, as an anomalous east-west, cold-warm surface temperature differential in the northern Indian Ocean brought torrential rainfall and flood disaster declarations by the U.S. Ambassadors to Somalia, Kenya, Ethiopia, Tanzania, and Djibouti.

While the failure of widespread, severe drought to materialize contrasted with public expectations, the more moderate, probabilistic predictions of the consensus outlook guidance issued prior to and midway through the 1997/98 season exhibited skill levels well above what would be expected by chance (Ward, Unganai, and Garanganga, 1998). The encouraging performance of the forecasting process notwithstanding, however, a bigger factor to consider is that drought fears reportedly caused some farmers to reduce the area planted and to devote less labor to agriculture generally, which may suppress food production and resulted in severe economic setbacks and heightened livelihood insecurity. A public backlash, generated when the worst failed to materialize, has implications for long-term acceptance and use of advance climate information.

An assessment of prospects for continuing progress in adapting to climate variability in southern Africa must take into account the overall trends during the 1990s, but particularly the 1997/98 experience. Clearly, the decade was one of transition from monitoring to prediction and from post-disaster relief to preparedness and prevention approaches. Yet the tone for the immediate future will be set by the most recent events.

Trends in the 1990s

Throughout the decade, awareness of climate impacts and the potential to improve their management grew for several reasons, including: (a) successive droughts and their negative socio-economic impacts, (b) sustained efforts by some national governments and international organizations and agencies to improve capacity to manage climate variability, and (c) the major 1997/98 El Niño that raised fears of yet another drought, catalyzing prevention and preparedness efforts and bringing them to public attention. Consequently there is now a wider understanding of causes of climate variability (e.g., El Niño) and the potential to forecast seasonal climate.

Prevention and preparedness strategies require acting earlier, however, and on less certain information. Experiences during the 1997/98 El Niño make it clear that additional work needs to be done towards improving forecasts, developing risk management strategies, and developing and reinforcing linkages and information and communication mechanisms between climate scientists and a broad array of better-identified users of climate information. In a 1998 post-season evaluation, USAID southern Africa Mission directors commented favorably on the unprecedented degree of preparedness and proactive planning in the face of the threat of impending drought. Rather than backsliding, efforts by professional climate scientists and other sectoral specialists in the region have redoubled to continue to refine the production and use of climate information to mitigate climate fluctuations and capitalize on favorable conditions, with significant support by donors.

The mixed legacy of 1997/98 notwithstanding, several factors favor continued progress in coming years towards improved management of climate variability in southern Africa (and elsewhere). In agriculturally dependent countries, there is an avid desire by farmers and other producers for advance information on the likely quality of the upcoming agricultural season. Southern African farmers closely monitor a variety of indicators, including plants and animals, for clues. Furthermore, even poor households have diverse risk management strategies involving adjustment of labor and resource allocation, planting strategies and the like (Gibberd et al., 1996).

Climate outlook forums have continued uninterrupted since Kadoma, organized by the Drought Monitoring Center in Harare and southern Africa national meteorological services. These forums include active participation by actual and potential users of climate information, who provide feedback to improve the usefulness of forecast products. Universities and research centers contribute to improving climate prediction techniques and educating users (Basher et al., 2001).

The 1997/98 experience focused attention on how climate information is transmitted to, and understood by, various user constituencies. One thrust is sectoral, in which efforts are being made to reach targeted users through sector-expert intermediaries (Eakin, 1998). A second is to develop specific messages designed to go directly to the end-user, or farmer. For example, the RANET project implemented by the African Center for Meteorological Applications in Development and the University of Oklahoma, funded by USAID, will work in Zambia and Mozambique to transmit climate information for household-level applications through wind-up radios in local languages.

Future Directions

The 1997/98 El Niño event generated an enormous amount of experience on applying advance climate information to reduce societal vulnerability to interannual variations in seasonal climate. Worldwide, an organized and systematic effort was made to publicize the event and accompanying risks (NOAA, 1999). Many nations in affected regions, including southern Africa, took the warnings seriously and took proactive measures.

Clearly, there is a widespread perception that such efforts can help societal adaptation to climate variation. While it is important to continue improving forecast capabilities, some of the biggest gains are likely to be realized by leveraging value from scientific information already available (Jarrell, 1999). Continued pursuit of preventive strategies will require a broader, more sophisticated engagement with a better defined user community that includes, above all, the general public.

Increasingly finely calibrated information reaching the hands of a few professionals is important, but even greater benefits to food security could be realized if existing knowledge were made more available to improve the ability of larger numbers of people to make even small adjustments at the household level based on anticipated climate conditions. Despite the potential value of improved forecasts and dissemination, forecasts are but one part of improved livelihood and sustainable development in many countries. Risk management decisions based on less certain advance information must, therefore also take into account many other factors affecting the circumstances of the user, such as resource availability, the range of alternative actions, and the consequences of potential outcomes. While there is value in continuing to work to improve international, regional and national-level preparedness for extreme climate events, a broad-based effort to communicate currently available information could contribute greatly to improved management of climate variation and disaster prevention at the grassroots.

References

Basher, R., Clark, C. Dilley, M. and Harrison, M. (eds) (2001), *Coping with the Climate: A Way Forward*, International Research Institute for Climate Prediction, New York.

Benson, C. and Clay, E. (1998), *The Impact of Drought on Sub-Saharan African Economies: A Preliminary Examination*, World Bank Technical Paper No. 401, World Bank, Washington, D.C.

Boudreau, T. (1997), Combining Risk Map Analysis and Climate Prediction: the Possible Effects of El Niño on Rural Households in Zimbabwe, Save the Children Fund-UK, London.

Callihan, D.M., Eriksen, J. H. and Herrick, A.B. (1994), *Famine Averted: The United States Government Response to the 1991/92 Southern Africa Drought*, Management Systems International, Washington, D.C.

Cane, M.A., Eshel, G. and Buckland, R.W. (1994), 'Forecasting Maize Yield in Zimbabwe with Eastern Equatorial Pacific Sea Surface Temperature', *Nature*, vol. 370, pp. 204-205.

Dilley, M. (1996), 'Diary of a Drought Year: Usefulness of Climate Forecasting Information for Drought Mitigation and Response from a Donor's Standpoint', *Workshop on Reducing Climate-Related Vulnerability in Southern Africa*, 1-4 October 1996, NOAA, Silver Spring, MD, pp. 141-148.

Dilley, M. and Heyman, B.N. (1995), 'ENSO and Disaster: Droughts, Floods and El Niño/Southern Oscillation Warm Events', *Disasters*, vol. 19, pp. 181-193.

Eakin, H. (1998), Using Seasonal Climate Forecasts in Farm Management, a Guide for Agricultural Extension Agents, Agritex Crop Production Branch, Harare.

Food and Agriculture Organization of the United Nations (FAO) (1999), *FAOSTAT StatisticalDatabase*, http://apps.fao.org/.

Gibberd, V., Rook, J., Sear, C., and Williams, J.B. (1996), *Drought Risk Management in Southern Africa: The Potential of Long Lead Climate Forecasts for Improved Drought Management*, Report to the Overseas Development Administration and the World Bank, Natural Resources Institute, University of Greenwich, Kent, U.K.

Glantz, M. (2001), *Lessons Learned from the 1997-98 El Niño: Once Burned, Twice Shy*, A UNEP/NCAR/UNU/WMO/ISDR Assessment, October 2000.

Glantz, M., Betsill, M. and Crandall, K. (1997), *Food Security in Southern Africa: Assessing the Use and Value of ENSO Information*, National Center for Atmospheric Research, Boulder, CO.

Halmrast-Sanchez, T., Dilley, M., Ashley, J., Hodgkin, J., Morse, T., Robles, A. and Kolwicz, C. L. (1995), *Southern Africa Drought Assessment*, 10-30 September 1995, U.S. Agency for International Development, Office of U. S. Foreign Disaster Assistance, Washington, D.C.

Halpert, M. S. and Ropelewski, C. F. (1992), 'Surface Temperature Patterns Associated with the Southern Oscillation', *Journal of Climatology*, vol. 5, pp. 577-593.

Jarrell, J. (1999), 'Guest Editorial: What Does the National Hurricane Center Need from Social Scientists?', *Weatherzine*, vol 15.

Jury, M.R. (1996), 'Malaria Forecasting Project', *Workshop on Reducing Climate-Related Vulnerability in Southern Africa*, 1-4 October 1996, Victoria Falls, NOAA, Silver Spring, MD, pp. 75-84.

Landman, W.A. and Mason, S.J. (1999), 'Change in the Association between Indian Ocean Sea-surface Temperatures and Summer Rainfall over South Africa and Namibia', *International Journal of Climatology*, vol. 19, pp. 1477-1492.

le Seur, D. and Sharp, B. (1996), 'Malaria Forecasting Project', *Workshop on Reducing Climate-Related Vulnerability in Southern Africa*, 1-4 October 1996, Victoria Falls, Zimbabwe, NOAA, Silver Spring, MD, pp. 65-74.

National Oceanic and Atmospheric Administration (NOAA) (1996), *Workshop on Reducing Climate-Related Vulnerability in Southern Africa*, 1-4 October 1996, Victoria Falls, NOAA, Silver Spring, MD.

National Oceanic and Atmospheric Administration (NOAA) (1999), An Experiment in the Application of Climate Forecasts: NOAA-OGP Activities Related to the 1997-98 El Niño Event, NOAA, Silver Spring, MD.

National Oceanic and Atmospheric Administration (NOAA), National Science Foundation, National Aeronautics and Space Administration, U.S. Department of Energy. (1996), *International Forum on Forecasting El Niño: Launching an International Research Institute*, NOAA, Silver Spring, MD.

Rook, J.M. (1996), 'Annex 1: Macro-level Benefits and Implication of Long Lead Climate Forecasting for Drought Risk Management', in V. Gibberd, J. Rook, C.B. Sear, and J.B. Williams, *Drought Risk Management in Southern Africa, vol 2: Technical Annexes*, Natural Resources Institute, University of Greenwich, Kent, UK, pp. 3-20.

Ropelewski, C.F. and Halpert, M.S. (1987), 'Global and Regional Scale Precipitation Patterns Associated with the El Niño/Southern Oscillation', *Monthly Weather Review*, vol. 115, pp. 1606-1626.

Thompson, A., Jenden, P. and Clay, E. (1998), Information, Risk and Disaster Preparedness: Responses to the 1997 El Niño Event, SOS Sahel International, London.

Thompson, C.B., in association with the Food Security Unit, Southern African Development Community (1993), *Drought Management Strategies in Southern Africa: From Relief through Rehabilitation to Vulnerability Reduction*, UNICEF Policy Monitoring Unit, Windhoek, Namibia.

U.S. Agency for International Development (USAID), Office of U.S. Foreign Disaster Assistance (OFDA). (1992a), *Southern Africa Drought Assessment*, 24 March -29 April 1992, USAID/OFDA, Washington, D.C.

U.S. Agency for International Development (USAID), Office of U.S. Foreign Disaster Assistance (OFDA). (1992b), *Annual Report FY1992*, USAID/OFDA, Washington, DC.

U.S. Agency for International Development (USAID), Office of U. S. Foreign Disaster Assistance (OFDA). (1996), *Disaster History*, USAID/OFDA, Washington, D.C.

Ward, M.N., Unganai, L.S. and Garanganga, B.J. (1998), 'Verification of the 1997/98 ENSARCOF Seasonal Rainfall Outlook Maps - Issues, Methodology and Results', in M. Harrison, (ed.), *Second Report of the ENRICH Southern Africa Regional Climate Outlook Forum to the European Commission*, September, 1998, United Kingdom Meteorological Office, U.K.

World Food Program (WFP), Southern Africa Regional Office. (1997), *An Initial Framework for Drought Contingency Planning for the World Food Programme*, World Food Program, Maputo, Mozambique.

3 National Responses to Seasonal Forecasts in 1997*

ANNE THOMSON

Introduction

Over the past years there has been considerable investment in the development of disaster preparedness and management systems in southern Africa. SADC has funded and facilitated training programs, and a number of agencies have supported the establishment of structures mandated to coordinate the preparedness for and response to natural and human disasters ranging from droughts and flooding to civil uprisings.

Food security issues have been a leading concern among governments and the donor community. Almost all countries in southern Africa have an Early Warning System (EWS) that predicts food production, evaluates trade flows, and then assesses the likely food situation at the national (and in some cases regional) level. Governments usually make assessments of likely production levels at various points in the growing season, starting at post-planting and extending through to early harvest assessments. Such evaluations may be combined with market price information to assess stresses and shortages in the food system. More sophisticated systems may even include household food security data from the post-harvest period. In some countries, government organizations have a specific mandate under exceptional circumstances (e.g., when food shortages are predicted) to import foods or impose trade restrictions. Such responses have been increasingly less frequent since market liberalization policies were adopted. Nevertheless, they can still be found, for example, in Tanzania, where the

---

* This chapter is based on work funded by the U.K.'s Department for International Development ESCOR fund. More detail can be found in Thomson, Jenden, and Clay, 1998.

Food Security Department still imposes export bans under extreme circumstances (Ministry of Agriculture and Co-operatives, 1997).

The main thrust of food security assistance is almost always to assess the need for intervention after there is evidence that the harvest has been poor. This could involve intervention at the national level, or delivery of food assistance to insecure households. The 1997/98 El Niño event posed an unprecedented challenge to governmental organizations and structures because the availability of seasonal climate forecasts raised the possibility of providing advice and taking actions before the production season started, thereby potentially reducing the possibility or extent of a poor harvest.

This chapter examines how national public sector organizations rose to this challenge in 1997. The study concentrates on actions taken by the governments of Namibia, Zambia, and Malawi prior to mid-November, when forecast information was available and the actual weather conditions for the growing season were still uncertain. Although responses extended to areas of water, energy, and health services, the focus of this chapter is limited to the agricultural sector. A number of questions are considered:

- What could governments reasonably have been expected to do in response to the seasonal climate forecast?
- What plans and actions did government organizations actually take?
- What constraints did government organizations face?
- Were their responses effective in hindsight, given the actual seasonal outcome?

Government Responses: The Options

In most southern African countries, the role of government has changed considerably over the past two decades (Sandbrook, 1990; Gallagher, 1994; Jeffries, 1993). Whereas previously the state was often a major provider of goods and services, and often intervened extensively in market operations, liberalization and an increased emphasis on the private sector has led to an acceptance that the state should redirect its focus to addressing market failures. This fact has to be taken into account when analyzing what governments might be expected to do in response to seasonal forecasts. In the 1960s and 1970s, when both food and input markets were heavily controlled by the state, an expectation of a poor growing season would have led to changes in the types of seed and inputs distributed through parastatal marketing agencies to reduce the risk of harvest failure by changing crop production patterns. In addition, public food stocks would have been built up to allow food to be released onto the market to stabilize food prices and prevent the development of shortages.

There is no way of knowing with certainty how effective these parastatal organizations would have been at reducing the risk of harvest failure given seasonal climate forecasts, as they were not available during the peak period of state responsibility for the food system. However, there is little evidence to indicate that parastatal organizations were particularly effective at either financing or implementing the necessary increases in imports or food stocks to stabilize and supply domestic cereal markets in times of harvest failure (Jones and Wickrema, 1999). In fact, a perceived inefficiency and lack of effectiveness was one of the factors behind the push towards liberalization of agricultural markets in the 1980s and 1990s.

In most SADC countries, the role of the state in the agricultural sector has been redefined to remove any elements of direct production or provision of services such as marketing. Government involvement has been largely restricted to areas where there is arguably an element of public good. This would include information provision and some elements of research and extension – areas where beneficiaries generally do not have the means to pay for the services themselves. Governments are usually responsible for coordinating and planning under these circumstances.

In this environment, the main function that governments would be expected to perform in response to a forecast for a poor growing season is to disseminate information, both about the forecast and about appropriate responses to reduce stakeholder risk. It is also reasonable to suppose that governments would undertake some form of contingency planning, either in terms of ensuring adequate food supply or preparing for an increase in the level of safety nets. These are fairly traditional responses to harvest failure (see von Braun, Teklu, and Webb, 1998). Climate forecasts would enable an extended preparation time for such responses. Elements of these possible responses were evident in Namibia, Zambia and Malawi, but they were implemented with varying degrees of success.

Actual Responses to the Forecasts

Information Dissemination

Information about the upcoming El Niño event in 1997/98 was available to governments and those who had access to the Internet and international media. National media in both radio and newspapers took up readily the potential consequences of what was projected to be the strongest El Niño on record. Within the governments studied, different ministries and units were active in spreading information to the public. Although the Met Offices were involved in public dissemination in all three countries, they took a particularly proactive role in Zambia, where an official statement was made as early as July 22, 1997. Zambian officials from the Met Office were heavily involved in government meetings, where they interpreted the seasonal forecast for officers in other ministries. In Namibia, officers from the Met Office were also heavily involved in inter-ministerial meetings, and they attended various farmer meetings to give information about the interpretation of the forecast. Such activities were less evident in Malawi.

Although governments did a reasonably good job of informing organizations and stakeholders based in the major cities about the seasonal forecasts, most of these agents had access to information from other sources, and were in any case likely to initiate requests for information from the government. There was significantly less information distributed to small-scale farmers, particularly in Zambia and Namibia.

In Zambia and Namibia, the channels for information flow to small-scale farmers are relatively limited. In both countries, small-scale farmers are much less organized than larger commercial farmers. Small farmer unions exist, but are poorly funded and often overburdened with requests to represent the views of the small farmer in a number of different government and international donor forums. These unions generally did not have the funds or capacity to distribute information to their members in the same way that commercial farmers unions did during this period. There were, however, exceptions. For example, in Namibia, the Caprivi Farmers Union organized their members and discussed the forecast and responses they should take.

The usual channel that governments use to communicate with small farmers is the extension service. Much has been written about the failings of the extension service in many parts of Africa (Purcell and Andersen,

1997). Many farmers have little contact with their extension officers, and therefore would be unlikely to receive timely information from them. However, in cases where contact was satisfactorily maintained, extension officers would not simply provide information about a weather forecast, but would also provide advice as to what farmers should do to reduce the risk from bad weather.

Risk Reduction Advice

Development of appropriate advice involves considerably more time and effort than simply disseminating a climate forecast. Translating a forecast into recommendations for appropriate actions requires an understanding of both climate impacts and mitigation options. Two of the three countries studied did provide advice to farmers. In the case of both Zambia and Malawi, agricultural extension services were used to deliver recommendations. In Namibia, the government intentionally refrained from providing such advice.

In Malawi, extension messages are prepared on an annual basis by the responsible department within the Ministry of Agriculture, and distributed through the extension network. Advice to concentrate on drought-resistant varieties was given to almost all Agriculture Development Districts (ADDs), although this advice was not directly linked to either El Niño or to the seasonal forecast in the official extension messages.

In Zambia, the technical committee of the National Early Warning System (NEWS) developed a comprehensive and well-presented leaflet advising farmers as to how to reduce drought-associated risks. The leaflet was written in English, with the intention of having it translated into local languages for easier access. However, the Ministry of Agriculture was unable to find funding to print the leaflet in significant numbers (possibly for bureaucratic reasons), even in English, and dissemination appears to have been limited to faxes to regional extension headquarters.

In Namibia, the Ministry of Agriculture took a very cautious approach to the forecast. When asked, they stated that their advice to farmers was to plant conservatively. A concern was expressed that if too much publicity were given to the possibility of crop failure, there was a risk that farmers would not plant at all, and instead simply rely on government assistance in the post-harvest period.

Contingency Planning

With the exception of Malawi, there was little evidence of coherent contingency planning for mitigation among governments. During fieldwork, researchers frequently encountered the attitude that a drought is not certain, therefore it is best to wait until something actually happens. This was particularly true where engaging in short-term preparedness would involve either considerable financial costs or political risks. There seemed to be two principal reasons for this wait-and-see attitude. First, at this stage there is little faith in the reliability of forecasts. Although large-scale commercial farmers in Namibia and South Africa are prepared to buy forecasts from proven forecast organizations, small-scale farmers have little experience of receiving forecasts on a timely basis. In general, they are at the end of a long and unreliable information chain. In addition, the wide area over which most seasonal forecasts are made means that, for an individual farmer in a country where there is considerable spatial variation in weather conditions, there is quite a good chance that the forecast will not be accurate for local conditions.

Second, the emphasis on probability in seasonal forecasting makes them even more suspect to many. Most of the public sector stakeholders affected by the 1997/98 El Niño predictions had limited experience in dealing with information presented in terms of probabilities and levels of risk. Much of the emphasis, particularly in terms of drought management, has instead been on improving the precision of estimates based on what has already happened. Where probabilities have been included in analysis and policy, past events have been used to calculate expected values based on average levels of risk.

Constraints to Government Responses

Financial

Although funding is almost always a potential constraint to new activities in developing countries, it was very seldom given as a major constraint in this study. This may have been in part because there was a strong interest in the donor community taking preventive action in cases where it would be potentially effective. For example, UNDP funded some initial steps towards contingency planning in Malawi. In the case of printing the advisory leaflet in Zambia, where financing was ostensibly the problem, it appears that internal bureaucratic problems may have contributed to the problem. There seemed to be a perception among some governments that there was a limit as to how much the government should be doing, particularly given the great risk associated with committing funds on the basis of a probabilistic climate forecast.

Lack of Existing Structures and Strategies

Where structures and strategies for disseminating information and advice existed, some flexibility was necessary in order to make use of them. Examples include the implicit incorporation of the climate forecast in Malawi's extension messages and the use of the NEWS technical committee in Zambia to develop an advisory leaflet. However, in the countries examined, the drought/disaster structures were either uncoordinated or in early stages of development and thus not capable of playing a leading role in promoting government actions in relation to the forecasts. In some cases, there was conflict as to whether drought was a matter for the Ministry of Agriculture or the disaster preparedness structures.

Perhaps most problematic is the nature of the EWSs, which are highly centralized and tend to work around a timetable that starts shortly after sowing and finishes after the final harvest assessment. Circulation lists for bulletins are heavily focused on government and donors, and few are sent to rural areas. Dissemination of information to the grassroots tends to be left to politicians, who are not necessarily neutral and objective.

Uncertainty

There are genuine problems as to how to respond to the level of uncertainty implied in the current format of seasonal climate forecasts. First, there is a general distrust of forecasts in some circles, often based on past experience. Second, even if the forecast is trusted, interpreting what it means is not simple. The implications of an increase in the probability of below average rainfall (e.g., from 33% to 45%) are not easy to understand in the absence of explanations or education. Such information could be better interpreted if it were translated into possible rainfall levels. Even then, the implications would differ according to whether a farmer normally plants maize or rootcrops, beans, or cabbage.

Finally, even if the forecasts are regarded as accurate and the implications are easily interpreted and understood, they remain probabilistic (i.e., expressed in terms of a probability within a given range of outcomes). Attitude to risk and the relative pay-offs from taking action versus doing nothing will affect what action is taken in response to these forecasts. Civil servants may be more risk-averse than farmers, and it is certain that their costs and returns to action will be different than for those facing farmers. It is also unlikely that the action they take in the face of uncertainty would be as immediate as those of other stakeholders. In short, the forecasts may be too uncertain to be of much use to farmers or even agricultural planners.[1]

Lessons Learned

What was the impact of the El Niño event on harvests in the three countries studied? In Namibia, coarse grain production was estimated to have fallen to 60% of its normal levels as a result of erratic rainfall and prolonged dry spells in March–April, particularly in the northeast. In Zambia, maize production was estimated at almost 50% of the normal level, partly because of flooding in the northern regions and near-drought conditions in the south. In Malawi, FAO estimated the 1998 harvest at 40% above average, as the result of abundant and widely distributed rains (FAO/GIEWS, 1998). However, it should be noted that the bumper crop in Malawi was also affected by the distribution of starter packs of seeds and fertilizer to small-scale farmers.

The observed climate results are consistent with the consensus forecast issued after the Kadoma meeting in 1997 with respect to the second half of the growing season. However, they were not consistent with the interpretation or emphasis placed on the forecasts by various public agencies responsible for information dissemination and contingency planning. As mentioned above, the agricultural extension messages for almost all ADDs in Malawi emphasized planting drought-tolerant crops. In Zambia, the drought pamphlet correctly indicated that poor rainfall was not expected in the north of the country, yet it did not even raise the possibility of flooding as a problem.

In hindsight, could governments have done more? In Malawi, the Ministry of Agriculture could perhaps have done less, and reduced the emphasis on drought-tolerant crops in inappropriate parts of the country. The advice given was extremely risk-averse, and could have imposed substantial costs on already poor farmers. Nevertheless, it is unlikely that the extension advice was acted on to any great extent. In Zambia, it would have been useful if the advisory leaflet that was developed had also been translated and disseminated to the appropriate regions in the south by the beginning of October. It is not clear whether there would have been sufficient planting material and seed available for farmers to act on the advice, though the Programme Against Malnutrition (PAM) was receiving supplies to assist farmers to do this.

In Namibia, it is more difficult to assess how government action could have been improved. As of the beginning of November, more effort could have been made to disseminate information through the extension service and to contact local farmer organizations directly. Where farmer organizations did have information, they acted to try to access appropriate seed, both domestically and from Zambia. However, there is little evidence as to how effective this strategy turned out to be.

Conclusions

With respect to seasonal forecasts in general, the two key areas seem to be information dissemination and what, if any, advice to provide along with the forecast. There are two sets of implications that can be gathered from the 1997 El Niño event: one relates to drought preparedness and the other to the more general issue of government responsibility with regard to seasonal forecasts.

With respect to drought preparedness, the most important lesson is that preparedness cannot be undertaken in three months, but must be a longer-term concern. Those who prepare best are those who have a structure developed that can be brought into play when the need arises. Information dissemination requires some kind of network, whether through radio, farmer unions, or extension services, where information flows from the center outwards, and not, as is the norm in many countries, from the grass roots level upwards. There is need for a better understanding than currently exists in many countries as to what the full impact of drought is likely to be.

There is also a need for a much greater understanding as to the implications of the forecasts. This is both in terms of how strong the forecast actually is and in terms of what the forecast means for agriculture. One implication is that more emphasis should be put on agro-meteorological research to develop a better understanding of how rainfall affects production levels within a country. Another is that extension advice and messages should provide more sophisticated predictions regarding rainfall patterns. However, given that many farmers appear to either have little contact with their extension workers, or feel that when they do, extension workers have insufficient understanding of the specific conditions of the locality, it may be that the government's prime responsibility must be to ensure good dissemination networks for seasonal forecasts, and leave farmers to respond as they feel appropriate.

Note

1   Many governments have recently faced the issue of how to present information and advice when there is an important element of risk and uncertainty involved (not only with respect to climate forecasts, but also with respect to food safety, disease, etc.). The UK government recently laid out some guidelines for using scientific advice in connection to the bovine spongiform encephalopathy (BSE) crisis in cattle (Department of Trade and Industry, 1997). The key issues include:
    * accessing the best available scientific advice;
    * involving at least some experts from other disciplines (not necessarily scientists) to ensure that the evidence is examined from a wide range of viewpoints;
    * maintaining transparency and openness in the process and in explaining the interpretation of scientific advice; and
    * the appropriateness of early communication with key interest groups.
    These may be useful principles for governments to bear in mind when disseminating information that is uncertain, such as climate forecasts. This is particularly the case where the forecast represents a potentially extreme situation, as was the case with the 1997/98 El Niño.

References

Department of Trade and Industry (DTI). (1997), *The Use of Scientific Advice in Policy Making*. London, U.K.

Gallagher, M. (1994), 'Government Spending in Africa: A Retrospective of the 1980s', *Journal of African Economies*, vol. 3, pp. 65-95.

Jeffries, R. (1993), 'The State, Structural Adjustment and Good Government in Africa', *Journal of Commonwealth Political Studies*, vol. 31, pp. 20-35.

Jones, S. and Wickrema, S. (1999), *Price Stabilisation Policies in the Context of Market Liberalisation*. Oxford Policy Management Briefing Note No. 3.

Ministry of Agriculture and Co-operatives, Agricultural Sector Management Project. (1997), *Food Security in Tanzania: Transport, Markets and Poverty Alleviation*, Ministry of Agriculture and Co-operatives, Dar es Salaam.

Purcell, D.L. and Andersen, J.R. (1997), Agricultural Extension and Research: Achievements and Problems in National Systems, World Bank, Washington D.C.

Sandbrook, R. (1990), 'Taming the African Leviathan', *World Policy Journal*, vol. 7, pp 673-701.

Thomson, A., Jenden, P. and Clay, E. (1998), Information, Risk and Disaster Preparedness: Responses to the 1997 El Niño Event, SOS Sahel, London.

von Braun, J., Teklu, T., and Webb, P. (1998), *Famine in Africa: Causes, Responses and Prevention*. John Hopkins University Press, Baltimore.

4 Forecasts and Farmers: Exploring the Limitations

ROGER BLENCH

Research addressed to questions framed by climate science is not necessarily useful to those whom climate affects. A climate forecast is useful to a particular recipient only if it is sufficiently skilful, timely and relevant to actions the recipient can take to make it possible to undertake behavioral changes that improve outcomes.

(Stern and Easterling, 1999, p. 37)

Introduction

Recent technical progress has increased the data available for producing seasonal climate forecasts, and the production of actual forecasts has expanded considerably. Seasonal forecasts are of potential significance to farmers and livestock producers, health professionals, economists, and those in the insurance industry (Stern and Easterling, 1999). In regions of the world that depend primarily on rainfed farming, foreknowledge of the likely pattern of precipitation could lead to substantial improvements in risk management strategies, as well as increased profits for larger-scale producers (Akong'a et al., 1988; Buckland, 1997).

Climate forecasts have indeed generated great interest in the arid and semi-arid regions of the world, where food-security crises recur with increasing regularity. If climatic conditions could be predicted to a degree that would make it possible to respond effectively in the agricultural sector, forecasts would potentially have a major impact on food security (Gibberd et al., 1996). Yet despite the considerable interest in and enthusiasm for seasonal climate forecasts, there is a considerable gap between the information that is provided by meteorological services and that which is useful to small-scale farmers (Joubert, 1995; Hulme, 1996; Blench, 1999b). The human and economic costs of poor information can be considerable; therefore strategies for expanding forecast use to small-scale farmers must be carefully developed.

This chapter reviews the problems associated with transforming forecasts into practical information for farmers, particularly small-scale farmers. First, rainfall characteristics that are important to farmers are discussed and discrepancies between scientific assessments and farmer evaluations of the forecasts are considered. Second, issues surrounding forecast dissemination are presented. The chapter then considers farmers' existing strategies for addressing risks, and the implications for future development of forecasts.

Forecast Characteristics and Testability

Seasonal climate forecasts provide probability-based predictions of the likelihood of certain rainfall conditions in the subsequent six months. They are constructed based on statistical analyses of rainfall, numerical modeling, and multivariate empirical approaches (Washington and Downing, 1999). For example, measurements of sea-surface temperatures in the Pacific enable forecasters to model the likely evolution of the Southern Oscillation in some detail (NOAA, 1994; Nicholson and Kim, 1997; Delecluse et al., 1998; Latif et al., 1998). This, in turn, makes it possible to produce probabilistic forecasts of rainfall conditions in many parts of the world, particularly those where ENSO influences climate variability.

Seasonal forecasts predict aggregate rainfall, and may provide a rough estimate of the start of the rainy season. However, they provide no information about key characteristics of interest to subsistence farmers, such as distribution and intensity of rainfall. Although crops cannot grow without a certain absolute amount of rainfall, the distribution of that rainfall is critical; any crops that experience too long of a gap between rains may die, regardless of the total seasonal precipitation. Similarly, rains above a certain intensity may cause flooding or otherwise damage crops, causing stunting or types of mildew. Indeed, from a farmer's perspective, it is important to consider how rainfall interacts with other types of parameters, such as the species and cultivars of crops, soil type and structure, and residual moisture.

In contrast to deterministic forecasts, which are either correct or incorrect, probability-based forecasts assign likelihoods to a range of possible outcomes. If the probabilistic forecast holds that one of three possible outcomes will occur in a given season, then no actual outcome can falsify that prediction. In other words, whatever happens can be assigned to one of the three probabilistic outcomes. Scientists evaluate the skill of forecasts by assessing the results in relation to the "climatology," where each category is given equal probability.

If past datasets and predictions are available, it should, in theory, be possible to test models by retrodiction or "backcasting" of known outcomes. In other words, if a comparable dataset exists for the previous ten years and predictions are created based on the dataset available for each year, then it should be possible to test the accuracy of probabilistic forecasts over time. However, in reality this is never the case because the quality and density of data points included in the datasets changes from year to year. Moreover, accelerating interest in this topic in recent years has tended to increase the amount of data collected and its geographical range. As a consequence, using backcasting to test present-day forecast methods is not possible. The models and methods used in forecasting are thus difficult to test, making the validity of climate forecasts challenging to assess.

It is also difficult to validate seasonal forecasts because of the coarse geographic scale they represent. Tercile forecasts are presented over very large areas, with one grid in southern Africa covering thousands of square kilometers. Yet rainfall in semi-arid Africa can vary substantially over a distance of only a few kilometers. Thus it is likely that the highest probability tercile will actually occur in some parts of the region. Even if aggregated regional rainfall statistics validate the prediction, the forecast may be of negative value to farm enterprises located in patches characterized by variant rainfall patterns.

The significance of forecast validation can be seen in the example of the 1997/98 seasonal forecasts for southern Africa. Early indicators showed that an El Niño was developing in the southern Pacific, and predictions of drought were made for southern Africa and Melanesia and disseminated via the media (e.g., BBC, 1997). On the basis of the forecast, combined with the great media attention given to ENSO, preparations for drought were made throughout Southern Africa. Although most of the predicted effects of the ENSO did occur worldwide, including drought in Melanesia, there were unpredicted floods in the Horn of Africa, and relatively normal rainfall in much of southern Africa. This resulted in considerable skepticism when the expected drought failed to materialize. The unexpected climate pattern in Africa is thought to have developed from the presence of a body of cold water in the Indian Ocean (Landman and Mason, 1999).

While many farmers considered the seasonal forecasts for southern Africa in 1997/98 to be misleading, the forecasts themselves were considered to be correct (Ward, Unganai, and Garanganga, 1998). This illustrates the disjunction between technical evaluations of forecasts and the evaluations of farmers responding to the forecasts disseminated by the meteorological or agricultural services. A forecast can be technically correct in that it meets statistical or other criteria, but useless or misleading to farmers because it encourages them to respond inappropriately.

Identifying and Reaching Users

Despite these constraints, climate forecasts can be useful in two ways: at national or regional levels for planning purposes (see Chapters 2 and 3), and in principle at the farm level. For example, in systems where a substantial proportion of seeds are bought by farmers every season, private companies and governments can balance their purchase of seed types to meet potential demand. Similarly, where agricultural insurance plays an important role in rural production systems, premiums can be calculated more precisely and, in principle at least, thereby benefit farmers. The direct benefits to rural smallholders, however, are far less transparent because they depend on developments in rural infrastructure and changes in systems for disseminating information that will vary from one government regime to another. Indeed, smallholder farmers' existing strategies presently have a very poor "fit" with centralized information structures, and this will only improve if greater understanding of these strategies percolates through to ministries, extension agencies, and other responsible bodies.

The usefulness of forecasts depends strongly on the characteristics of potential users. At present, the disseminated outputs have tended to focus on a rather oversimplified dichotomy between large-scale commercial farmers and small-scale subsistence farmers. However, many farmers do not actually fit into either of these two categories, as illustrated by emerging farmers in South Africa (Vogel, 2000). Initial studies of the impact of forecasts on users suggest that a richer user-profile is required if the forecasts are to be beneficial to managers of farm enterprises. Some of the key parameters that have been identified are shown in Figure 4.1.

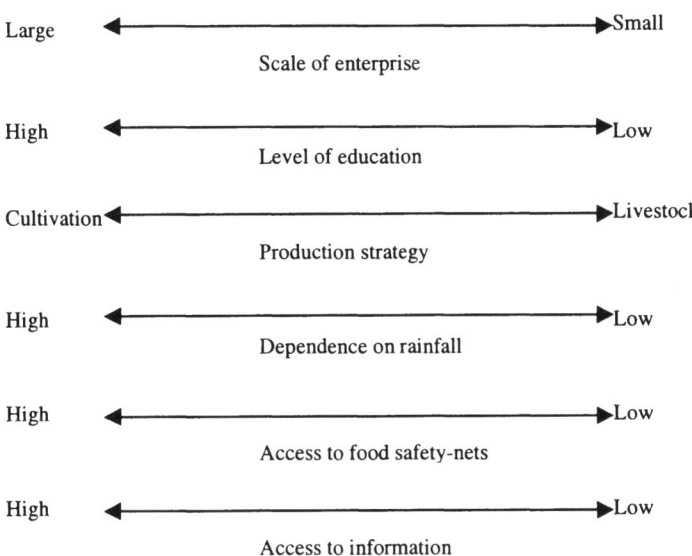

Figure 4.1   Parameters to characterize forecast users

An expanded classification of forecast users implies that the process of intermediation between the producers of technical forecast output (e.g., the weather bureaus and the Regional Outlook Forums) and the consumers (e.g., the farmers and pastoralists) will need to be increased substantially. In fact, it is likely that each group will have to be targeted differently in order to successfully disseminate seasonal forecasts. Assuming governments or their intermediaries have a confident message for crop and livestock producers, the problem of reaching farmers remains a challenge. A number of difficulties with current forecast formats may arise when disseminating forecasts. Dissemination of probabilistic forecasts in an understandable manner and in a form that is useful to farmers is unlikely without training and other support.

Agricultural extension services are often considered the most obvious channels for disseminating forecasts. However, if the experience of both NGOs and international agencies since 1960 has shown anything clearly, it is the problematic nature of extension services (Gautam, 2000; Christoplos, Farrington, and Kidd, 2001). In the colonial era, when the administration had rather clear economic goals, such as the production of cotton or groundnuts, and were not above using compulsion, agricultural extension was often effective within a limited sphere. Even today, when a cash crop such as tobacco is purchased by a well-established company, input supply to smallholders is usually effective. But improving the production and marketing of staples by strengthening government services has only been of limited value, and the general pattern seems to be one of deterioration.

Another aspect of the extension/dissemination debate concerns the evolution of trust between farmers and government extension agents (Blench and Marriage, 1998). Even prior to the advent of seasonal forecasts, indications in many parts of semi-arid Africa point to a collapse of trust between farmers and extension agents. Agents are often seen as being out of touch with the realities of agricultural life, recommending specific quantities of fertilizer applications to producers who cannot afford fertilizer at all. Unless they are integrated with other aspects of infrastructure and input supply, there is a clear risk of climate forecasts simply becoming categorized as another class of useless information. For example, a prediction of drought could be linked to increased availability of appropriate drought tolerant seeds. The dissemination of forecasts alone, however, is likely to only exacerbate the distrust that already exists.

An alternative dissemination strategy is to make use of farmers' existing information networks. Farmers and pastoralists maintain networks to collate information about crop varieties, input supplies, rainfall and government activity, and they are adept at integrating such information into their cropping or livestock production strategies. Rather than trying to interpret information, recommend agricultural strategies, and communicate such strategies through extension agents, the direct broadcasting of forecasts (for example, by radio) may be more effective than presenting farmers with crop recommendations. Only a limited number of farmers will take climatic predictions into account, but for those that do succeed, food security should gradually improve.

Existing Strategies for Addressing Risks

The technical aspects of climate forecasting are highly advanced compared to the mechanisms for making practical use of the results, particularly among small-scale and subsistence farmers. To integrate forecasts with national and local agricultural strategies, it is important to understand the basis on which farmers make decisions in semi-arid regions, as well as the context in which these decisions are made.

In a typical model, if farmers assume that the forecasts are correct, they would be expected to adopt a strategy reflecting the most probable outcome for that year (for example, below normal rainfall). But in the real world, no rational producer would gamble all of his or her resources on a single strategy, regardless of the perceived quality of information supplied. Farms are enterprises with limited resources of labor, cash, and land. A single year when total resources are wrongly directed could devastate the enterprise.

In addition, climatic and environmental factors represent only one aspect of overall production strategies. Farmers must also account for availability and cost of labor, seeds, pesticides and cultivation, as well the likelihood of receiving food-aid if they are wrong.[1] As a result, a response to climate that protects against losses, even at the expense of losing the opportunity to gain significant profits, assures the long-term continuity of an enterprise.

Although responses to the wide spectrum of risks vary between countries, communities, and individuals, African farmers have historically spread risks by managing diversity (Hardon and de Boef, 1994; Eyzaguirre and Iwanaga, 1996; van Oosterhout, 1996). Prior to the introduction of maize, the principal African cereals were sorghum, millet, rice and finger-millet, with the addition of fonio and iburu in West Africa and tef in Ethiopia. All of these cereals were notable for their phenotypic diversity, and large reservoirs of cultivars were maintained by individual communities. Farmers tended to plant both a mix of cultivars and a mix of species, with the exact composition reflecting both the date of the first rains and the period until the rains "set in" (Ellis and Galvin, 1994).[2] Moreover, in many areas water availability can be manipulated throughout the season, both by restructuring the pattern of ridges or beds and by hand-irrigation.

Seasonal forecasts are not always ideally suited to the way farmers actually farm in much of Africa. Different agricultural strategies are required for "normal" and low-rainfall years, and inaccurate predictions can have serious economic impacts on farmers. Reliance on probabilistic forecasts becomes riskier as a farmer's investments increase. A large-scale farmer, for example, must purchase hybrid seed of certain varieties in advance of planting. In contrast, a subsistence farmer often chooses seeds from a stock of stored cultivars, and plants them in the light of actual, rather than potential rain. Farmers may acknowledge the forecast, but they may not always be able to pursue their production strategy of choice.

A small-scale farmer invests a high proportion of labor and a low proportion of cash every year. The relative accessibility of arable land implies low transit and transaction costs incurred in repeatedly changing planting or tillage. Thus farming decisions in semi-arid areas can be characterized as having a dynamic relationship to rainfall, with the flexibility to change crops, mixtures, and soil management through the period of precipitation in response to small variations in climate. This is possible because of the small scale of farm size, labor pool, and cash.

Larger farms find it more difficult to follow such a dynamic strategy; seeds are purchased once, machinery hired once and labor paid at agreed rates. Such enterprises must make decisions once during an agricultural season, and stick with them. The assumption is that cash reserves, agricultural insurance or government fallback systems enable such enterprises to survive bad years.[3] These "single strategies" can be described as deterministic, where a single prediction prior to cultivation predetermines irreversible decisions throughout the season. Figure 4.2 shows a schematic representation of the contrast between these "single strategy" and "dynamic" behavior models.

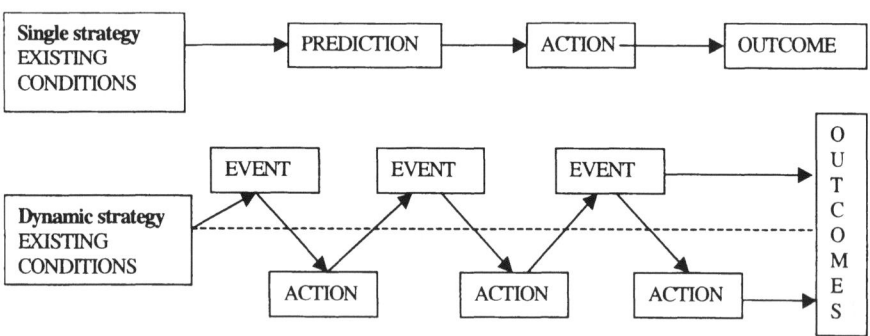

Figure 4.2   The contrast of single strategy and dynamic approaches

Implications for Policy

The intermediation process between users and producers of forecast information in African countries should be considered within the context, cultural background, and experiences of end-users. Forecasts may, for example, compete with indigenous knowledge systems, making their uptake initially slow. Farmers who depend on rainfed cultivation usually develop complex cultural models of weather, and may cite local predictors of seasonal climate (e.g., Pepin, 1996; Roncoli, Ingram, and Kirshen, 2002). Retrospective interviews often highlight indicators noted by farmers as prefiguring the weather pattern they in fact experienced.[4] Understanding the perception and categorization of weather and climate in different cultures is essential to the process of making meteorological information useful.

It is considered a truism that farmers who depend on rainfed agriculture across the world have evolved complex risk-aversion strategies. Yet farmers are no more immune to fashion than are other sectors of the economy, and they do not always follow a strictly agronomic rationale when they replace hardy "traditional" crops with "modern" varieties. Furthermore, the category of risk now includes changing government policies, varyingly effective infrastructure, the ebb and flow of aid agencies, NGOs and emergency relief programs, and the impact of the media.

The record of African governments' responses to predictions of drought in the policy arena has not been encouraging (Magadza, 1996). Nonetheless, since the mid-1990s very significant changes have occurred. Meteorological and climatic data have become more freely available, and meteorologists have accepted that they bear some responsibility in communicating its implications, as evident in the emerging exchanges of information between users and producers of climate forecasts during the various Regional Climate Outlook Forums. In turn, governments have adopted the idea that climate forecasts should inform policy over decades rather than single years, and that they can serve as an essential tool for planning. Moreover, the notion that forecasts have implications across many sectors of the economy and affect many types of players is gradually spreading, as is the notion that extreme weather events demand both regional planning and a regional response. Often constraints to action are financial, however, rather than a failure of political will.

Nonetheless, the impact of climate forecasts on small farmers is less evident. As argued in this chapter, there are several reasons for this: the characteristics of the forecasts themselves; the general weakness of extension systems; and a fundamental misunderstanding of farmers' decision-making processes. Probabilistic forecasts are adapted to larger-scale enterprises with deterministic decision-making and cash reserves or other safety nets. Making seasonal climate forecasts useful to small-scale farmers will involve the development of an integrated approach to infrastructure and input supply, as well as a thoughtful implementation of dissemination and translation strategies.

Notes

1   Food-aid, paradoxically, compels farmers to take more risks because the gains are high if they succeed and the state picks up the losses if they fail (e.g., Blench 1999a; Tripp 2001).

2   It is not unusual for farmers in West Africa to plant two or three times at the beginning of the rains, varying the cultivar mix each time, or even to uproot seedlings when it becomes clear they are not appropriate for the observed rainfall.

3   The same is true with crashes in livestock numbers due to drought or disease; pastoralists tend to be wiped out, but livestock farmers are compensated by the state (as in the recent foot and mouth epidemic).

4   The existence of ethno-meteorological models has sometimes also led to rather inappropriate reification of this type of knowledge by researchers, although farmers themselves seem to place only limited trust in these predictors. This can be illustrated by the example of the Bedu nomads in the Badia desert of Jordan, where both nomads and farmers estimate rainfall indirectly from the state of the vegetation. There is no doubt that vegetation cover is declining, so the Bedu are convinced that rainfall is decreasing throughout the rangelands (Blench, 1998). Vegetation degradation was, however, primarily due to poor management practices – in particular uprooting whole shrubs for firewood – rather than changes in rainfall. Indeed, records show that levels of precipitation had hardly varied from the 1920s when statistics were first collected.

References

Akong'a, J., Downing, T.E., Konijn, N.T., Mungai, D.N., Muturi, H.R. and Potter, H.L. (1988), 'The Effects of Climatic Variations on Agriculture in Central and Eastern Kenya', in M.L. Parry, T.R. Carter, and N.T. Konijn (eds), *The Impact of Climatic Variations on Agriculture, Assessments in Semi-Arid Regions*, Kluwer Academic Press, Dordrecht, Netherlands, pp. 123-270.

BBC. (1997), 'From our Own Correspondent: El Niño Special Edition', Transmitted November 1997.

Blench, R.M. (1998), 'Rangeland Degradation and Socio-economic Changes among the Bedu of Jordan: Results of the 1995 IFAD Survey', in V.R. Squires and A.E. Sidahmed (eds), *Drylands: Sustainable Use of Rangelands in the Twenty-first Century*, IFAD, Rome, pp. 397-420.

Blench, R.M. (1999a), 'Agriculture and the Environment in Northeastern Ghana: a Comparison of High and Medium Population Density Areas', in R.M. Blench (ed.), *Natural Resource Management and Socio-economic Factors in Ghana*, Overseas Development Institute, London, U.K, pp. 21-43.

Blench, R.M. (1999b), *Seasonal Climatic Forecasting: Who Can Use it and How Should it be Disseminated?* Natural Resource Briefing Paper 47, Overseas Development Institute, London, U.K., http://www.oneworld.org/odi/nrp/47.html.

Blench, R.M. and Marriage, Z. (1998), *Climatic Uncertainty and Natural Resource Policy: What Should the Role of Government Be?* Natural Resource Briefing Paper 31, Overseas Development Institute, London, http://www.oneworld.org/odi/nrp /31.html

Buckland, R.W. (1997), 'Implications of Climatic Variability for Food Security in the Southern African Development Community (SADC)', *Internet Journal for African Studies*, Issue 2, March 1997. http://www.brad.ac.uk/research/ijas/ijasno2/ijasno2.htm

Christoplos, I. Farrington, J. and Kidd, A. (2001), *Extension, Poverty and Vulnerability: Inception Report of a Study for the Neuchâtel Initiative*, ODI Working Paper 144, ODI, London.

Delecluse, O., Davey, M.K., Kitamura, Y., Philander, S.G.H., Suarez M. and Bengttson, L. (1998), 'Coupled General Circulation Modelling of the Tropical Pacific', *Journal of Geophysical Research*, vol. C7, pp. 14,357-14,373.

Ellis, J.E. and Galvin, K.A. (1994), 'Climate Patterns and Land Use Practices in the Dry Zones of East and West Africa', *Bioscience*, vol. 44(5), pp. 340-349.

Eyzaguirre, P. and Iwanaga, M. (1996), 'Farmers' Contribution to Maintaining Genetic Diversity in Crops and its Role within the Total Genetic Resources System', in P. Eyzaguirre and M. Iwanaga (eds), *Participatory Plant Breeding*, IPGRI. Rome, pp. 9-18.

Gautam, M. (2000), *Agricultural Extension: the Kenya Experience*, The World Bank, Washington, D.C.

Gibberd, V., Rook, J., Sear, C.B. and Williams, J.B. (1996), *Drought Risk Management in Southern Africa: the Potential of Long Lead Climate Forecasts for Improved Drought Management*, Natural Resources Institute, Chatham Maritime, U.K.

Hardon, J.J. and de Boef, W.S. (1994), 'Local Management and Use of Plant Genetic Resources', in A. Putter (ed.), *Safeguarding the Genetic Basis of Africa's Traditional Crops*. CTA, Netherlands and IPGRI, Rome, pp. 115-126.

Hulme, M. (ed.) (1996), *Climate Change and Southern Africa*. Report to WWF International by the Climatic Research Unit, UEA, Norwich, U.K.

Joubert, A.M. (1995), 'Simulations of Southern African Climate by Early-generation General Circulation Models', *South African Journal of Science*, vol. 91, pp. 85-91.

Landman, W.A. and Mason, S.J. (1999), 'Change in the Association between Indian Ocean Sea-surface Temperatures and Summer Rainfall over South Africa and Namibia', *International Journal of Climatology*, vol. 19, pp. 1477-1492.

Latif, M. et al. (1998), 'A Review of the Predictability and Prediction of ENSO', *Journal of Geophysical Research*, vol. C7, pp. 14,375-14,393.

Magadza, C.H.D. (1996), 'Climate Change: Some likely Multiple Impacts in Southern Africa', in T.E. Downing, (ed.), *Climate Change and World Food Security*, Springer-Verlag, Heidelberg, pp. 449-483.

National Oceanic and Atmospheric Administration (NOAA). (1994), *El Niño and Climate Prediction*. Reports to the Nation, 3, NOAA/OGP, Washington, D.C.

Nicholson, S.E. and Kim, E. (1997), 'The relationship of the El Niño Southern Oscillation to African Rainfall', *International Journal of Climatology*, vol. 17, pp. 117-135.

Pepin, N. (1996), 'Indigenous Knowledge Concerning Weather: The Example of Lesotho', *Weather*, vol. 51 No7, pp. 242-248.

Roncoli, C., Ingram, K. and Kirshen, P. (2002), 'Reading the Rains: Local Knowledge and Rainfall Forecasting in Burkina Faso', *Society and Natural Resources* vol. 15, pp. 411-430.

Stern, P.C. and W.E. Easterling (eds) (1999), *Making Climate Forecasts Matter*, National Academy Press, Washington, D.C.

Tripp, R. (2001), *Seed Provision and Agricultural Development*, ODI, London and James Currey, Oxford, U.K.

van Oosterhout, S. (1996), 'What Does In Situ Conservation Mean in the Life of Small-scale Farmers? Examples from Zimbabwe's Communal Areas', in L. Sperling and M. Loevinsohn (eds), *Using Diversity: Enhancing and Maintaining Genetic Resources On-farm*, IDRC, New Delhi, pp. 35-52.

Ward, M.N., Unganai, L.S. and Garanganga, B.J. (1998), 'Verification of the 1997/98 ENSARCOF Seasonal Rainfall Outlook Maps - Issues, Methodology and Results', in M. Harrison, (ed.), *Second Report of the ENRICH Southern Africa Regional Climate Outlook Forum to the European Commission*, September 1998, United Kingdom Meteorological Office, U.K.

Washington, R. and Downing, T.E. (1999), 'Seasonal Forecasting of African Rainfall: Prediction, Responses and Household Food Security', *The Geographical Journal*, vol. 165, pp. 255-274.

Vogel, C.H. (2000), 'Usable Science: An Assessment of Long-term Seasonal Forecasts amongst Farmers in Rural Areas of South Africa', *South African Geographical Journal*, vol. 82, pp. 107-116.

PART II:
CASE STUDIES OF
USER RESPONSES

5 The Use of Seasonal Forecasts by Livestock Farmers in South Africa

JERRY HUDSON AND COLEEN VOGEL

Introduction

Drought is a recurrent phenomenon in southern Africa that requires coping and adaptation in several sectors, including the agricultural sector. In southern Africa droughts are associated with sporadic rainfall that varies across time and space. Some droughts are localized while others are widespread; some mainly affect grass-production while others influence crop production (Hulley, 1980; NCAR, 1985; Wilhite, 2000). Coupled to this dynamically changing physical environment is the changing socio-economic agricultural context in which farmers and households operate. Changes in drought-subsidy schemes, land reform changes, and rising production costs are some of the socio-economic realities that confront farmers in their daily decision-making environment (Van Zyl, Biswanger and Kirsten, 1996; Simbi, 1998). To respond flexibly in dynamic physical and social environments, farmers must be able to draw on a wide range of coping and adaptation strategies. One possible way in which farmers could increase their adaptive capacity is through the use of seasonal forecasts.

In this chapter, we present a study that explores opportunistic management when drought conditions are either forecasted or actually occur for livestock farmers in the western parts of the North-West Province of South Africa. Specific details about drought management strategies used by farmers in the southern Kalahari are scarce, as are drought management responses between commercial and communal farmers living in this area (Freeman, 1984; Vogel, 1993, 1994). The role and uptake of seasonal forecasts as a means of adapting to climate variability is explored. Responses to minimize the effects of short- and long-term drought are illustrated by two different approaches: commercial farmers tend to decrease herd size and maintain higher average weight on their livestock, while communal farmers tend to maintain their herd size and have lower weight per animal.

Insight into the diversity of complex drought-coping strategies requires an understanding of the interaction of environmental and socio-economic factors within which management strategies are formulated in the region (e.g., Vogel, 2000; Patt, 2001; Phillips, Makaudza, Unganai, 2001). Results of this study should aid in the targeting and tailoring of forecasts for varied user groups. Data presented here are based on detailed field interviews of 25 commercial and 35 communal farmers. The farmers were interviewed during the winter and early spring of 1999.

Background

The Republic of South Africa is geopolitically and administratively divided into nine provinces, which are further divided into districts. The North-West Province (Figure 5.1) contains 28 districts, 5 of which comprise the western part of the province and contain just under half of the province's land. These districts, the focus of the study, are shown below: Vryburg One, Vryburg Two, Ganyesa, Kudumane, and Taung. The five western districts of the North-West Province generally lack sufficient rainfall, surface water and fertile soils to make crop production a viable enterprise (Leppan and Bosman 1923; Vogel 1994; Cowling, Richardson, and Pierce (1997) and livestock production is consequently a significant activity.

The Kalahari is sometimes erroneously called a desert because of its high average summer temperatures and lack of surface water. The Kalahari, however, has abundant vegetation, mostly in the form of shrubs and grasses and is more correctly described as a savanna environment (Acocks, 1975). Winters are relatively warm sometimes punctuated with cold fronts and lower temperatures that typically last less than a week. Early morning frost is not uncommon in the winter with temperatures above freezing during winter days. On clear summer days, temperatures rise into the mid to upper 40s ($^\circ$C). There is usually no precipitation from late autumn through winter, and well into spring. Farmers therefore typically know by the end of the summer growing season whether or not they will have ample forage to last through the winter and into the next spring. Although dry winter weather is very reliable and predictable, summer rains start and end unpredictably (usually starting in late September or early October and ending in April or May). Summer rains typically contribute to annual rainfall between 350 to 500 mm.

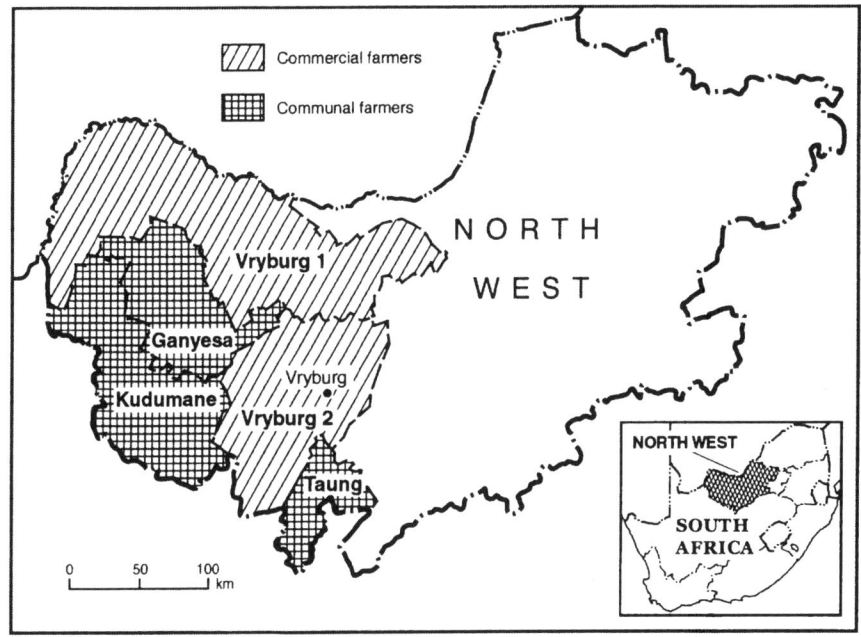

Figure 5.1 The five western districts of the North-West Province, RSA

Throughout the summer's wet season, rainfall is highly variable, both temporally and spatially. The 1999/2000 growing season, for example, exemplified both the temporal and spatial variability of rainfall patterns in the southern Kalahari. The season started with drought conditions because of the late onset of the springtime rains and lower amounts of spring rainfall than normal. The first rain of this season in the western part of the North-West Province fell fairly late in the spring (October 2, 1999), but was of sufficient amount to initiate substantial grass growth. The following rains, some two weeks later, were enough to sustain grass shoots until more frequent rains started in early November. Even though the 1999/2000 season started with a drought, unusually high amounts of rain came late in the summer and autumn, making this the wettest growing season in over 50 years.

Rainfall is, as indicated earlier, not only variable in time but also across space. Some farmers, at the beginning of a rainy season, enjoy several substantial rains, while neighbors located a few kilometers away might still be waiting for the first rainfall of the season. In the growing season of 1999/2000, for example, summer rains started early in October in some areas. By the first week in December, however, some farms had still not received any rain. This variability in the timing of rainfall can compound the set of options that farmers have. Having sketched the highly variable climate of the area, attention now shifts to examine the vegetation and grasslands.

The Vryburg area is located on the southern edge of the Kalahari. This land is characterized as an arid to semi-arid environment (Acocks, 1975), and is generally unsuitable for dryland crop production because of limitations in climate, soils, or terrain (Leppan and Bosman, 1923; Vogel, 1994). The diversity of grasses in the southern Kalahari is large, with several perennial and annual species (Roberts and Fourie, 1989). This diversity contributes to the resilience of the rangeland to periods of insufficient or inconsistent rainfall, and overgrazing. Land that appears severely over-grazed, for example, can be used as pasture just two weeks after the first substantial rains without causing further degradation of the pasture.[1] Although semi-arid pasture grasses are higher in nutrients and minerals than grasses from wetter climates, research by the South African Department of Agriculture has shown that animals thrive much better when given supplements, such as protein, salt, and vitamins (Mans, 2000). All of the commercial and communal farmers interviewed in this study routinely give their animals some dietary supplements for nutrition and health throughout the year but provide no supplementary fodder. Supplementary fodder is never given to animals, except in times of severe drought.

Living with Drought in the Southern Kalahari

There is no universal definition of drought, one definition generally accepted is that drought occurs when rainfall is 75% or less than the average rainfall (Laing, 1992). Droughts occur in a long-term pattern of climatic oscillations with a cycle of about 20 years. This periodicity is punctuated by droughts in the 1930s, 1950s, 1970s, and 1990s and wetter seasons in the 1940s, 1960s and 1980s (Tyson, 1986). These oscillating patterns of weather only show the general trend of yearly rainfall during the cycle. It is important to realize that severe droughts may occur during wetter parts of the oscillation, and conversely, wet seasons may occur during the drier half cycle of this long-term oscillation. The length of drought in South Africa is often unpredictable, and can last over a span of years.

The weather patterns identified above indicate that two forms of drought exist for farmers. The first type occurs when insufficient rains come during the rainy season to produce enough biomass for animals to sustain them from late spring, through the dry winter months, and into the spring-time growing season. The second type occurs when spring and summer rains come later than normal and thus farmers must either wait out the dry spell or implement drought mitigation strategies. Thus, the first form of drought usually arises from the problem of rainfall quantity, while the second type arises from the timing of rainfall.

Southern Africa, as indicated earlier, is prone to drought conditions, and people in the region are faced with the probability of any year being a drought year (Bruwer, 1989). In fact, drought is considered to be a natural event and drought preparedness is a normal part of management strategies. Farmers in the five western districts of the North-West Province typically expect a fairly severe drought after three or four years of plentiful rainfall. In the eastern districts of this province, where crop farming is the means of livelihood, droughts have a much more drastic effect on people's food security because crops are more susceptible to drought, especially short-term drought, than animals.

For many people in the world, the idea of drought brings forth images of dusty landscapes with dried carcasses of animals that have succumbed to the combined effects of thirst and intense heat. In the southern Kalahari these images seldom become real and occur in only the severest of droughts since animals rely on water from boreholes (wells) throughout the entire year. Even during the most severe droughts, animals in the five western districts tend to have sufficient water, but may slowly starve to death because of decreasing amounts of fodder. Drought in these parts would thus be described as a condition of insufficient rainfall to produce range fodder for winter and early spring seasons.

Droughts and El Niño

El Niño affects weather patterns on a global scale, and has been associated with floods and droughts on all continents of the world (Glantz, 1996; Changnon, 2000). There is a correlation between El Niño and many of the more severe droughts that have occurred in southern Africa (e.g., Mason and Jury, 1997; Lindesay, 1998). This is exemplified by the El Niño-driven drought that occurred in the summer of 1992, when less than half of the normal rainfall occurred. This drought was considered as being the worst in southern Africa since the beginning of the century (Harsch, 1992). An estimated 243,000 head of cattle and 101,000 head of small stock were lost in the three communal districts of the western North-West Province because of mortality in the 1992 drought (Rwelamira, 1997).[2] In this year, animal prices decreased dramatically due to the poor condition of animals. The animal market was saturated as farmers tried to sell their livestock before their animals became weak and sick, or starved to death. Thus, almost all farmers lost money as they were forced to sell their livestock at substantially reduced prices. Economic devastation resulted because livestock is a major asset for most rural farmers in South Africa.

Farmers in the study area rely on groundwater pumped by wind power for virtually all domestic and livestock needs and water availability is generally not a problem. Commercial farmers typically have numerous boreholes or pipe water to most of their rotational camps. Most commercial farmers have diesel or electrical backup power for their boreholes in the event of poor winds, and a few boreholes in communal villages also have electric or diesel backup power. In contrast to commercial farmers, many communal farmers report that water availability can become a serious short-term problem when winds are insufficient to operate the windmills that pump water for stock tanks.

This study also revealed that insufficient water in some communal areas is an ongoing constraining factor in the economic security of communal farmers. The scarcity of boreholes in communal areas means that as the winter progresses, animals are forced to forage further and further from their water supply. In the more remote areas of the study, animals can reach their distance limit of daily travel to water before spring rains replenish grasses closer to their water supply. The result is that animals can slowly starve because ample fodder may be just out of their reach.

Quality of water also varies a great deal within the region, but it is usually good enough for human and animal consumption. Water quality problems include unusually high lime and salt concentrations; and particularly in communal areas, water is sometimes contaminated by animal waste. Only in a very few localized areas is water not suitable for human consumption. Of the sixty farmers interviewed, all but four (three communal and one commercial) reported that water quality is never a problem for their animals.[3]

Farming in the Western Part of the North-West Province

Since land in the southern Kalahari is generally unsuitable for dryland crop production, the majority of people living in this semi-arid region farm livestock (mainly cattle, sheep, and goats) as their primary economic base. A substantial portion of the South African livestock industry is located within this region, and Vryburg, with a population of about 8,000, is the largest town in the area (Figure 5.1). Located near the center of the Vryburg Two district, Vryburg claims to have the largest cattle market in the southern hemisphere, and up to 250,000 cattle are sold annually, mainly by commercial farmers (Vryburg Development Center, 1999).

In the two Vryburg districts, white commercial farmers own large widely-spaced, western-style farms that are typically greater than 2500 ha and mainly used for cattle production. Farms in these two districts have been surveyed and are well-delineated with fences. Farmers typically have large farm houses, several outbuildings, and provide modest permanent housing for their workers and the worker's families. The vast majority of commercial farmers rotate animals using the camp system, in which animals are kept in pastures for a limited time to allow for controlled grazing and maximize grass production. About 70% of these farmers are Afrikaners who are descendants of Dutch settlers. The other 30% are of British descent. There are no known black landowners in these two districts.

The three other districts in the western part of the North-West Province, Ganyesa, Kudumane, and Taung, are part of the former Bophuthatswana

Homeland, which later became the Bophuthatswana Republic. Homelands were formed in the 1950s as a political adjustment under the nation's apartheid rule and were incorporated into the Republic of South Africa in 1994 with the abolishment of apartheid (Bromley, 1995). No known white farmers or landowners live in these three districts. These districts are inhabited by people of the Tswana tribes, who generally live in villages headed by a chief. Most Tswana people work at "subsistence" level animal farming in or around their village. Typically, animals are individually owned at the household level and are grazed upon communal pasture lands which have poorly defined boundaries. Fences are virtually non-existent and access to pasture areas may be controlled by the village chief, village council, or complex social norms. In some cases, the chief may allocate individual grazing rights to select pastures, in other cases there is no restriction on an individual farmer's grazing rights. Grazing land is shared not only by members of one village, but in many instances by people from two or more villages. Many communal farmers have electricity, but indoor plumbing is virtually nonexistent. Villages get water from one or more community wind-driven or diesel-powered boreholes. Household water is carried from a borehole to the home, typically in plastic jugs hauled by adolescent boys using wheelbarrows. It is common for animals and humans to share water from the same borehole.

Traditional socio-economic systems in all communal areas have been replaced by reliance on the market economy at the household level, and old customs of egalitarian sharing seem to have virtually disappeared. Typically, communal farmers support a large number of people in their nuclear and extended family and live at or just above the "subsistence" level. Many land tenure systems are used by communal farmers. The "4-40" system (4 farmers limited to 40 ha of pasture) established by the Bophuthatswana homeland government is still used by a small number of farmers. Other farmers are involved in the South African Development Test (SADT) Farm program, in which pasture lands were set aside for exclusive use by communal farmers. Under this homesteading program a farmer is allowed to raise animals on a piece of land for seven years, and after that time may apply for title to the land - providing the farmer has been able to make a profit and maintain the land in a "sustainable manner".

Living with Drought: Coping and Adaptation

As indicated above, drought is a recurrent phenomena in the southern Kalahari to which farmers must continually adjust their livestock management practices. The certainty of seasonal variability means that farmers have devised diverse livestock management strategies to minimize anticipated effects of likely future drought conditions. It is common for one drought year to follow another, and occasionally several years of sequential drought occur. Most commercial farmers think they can cope with droughts lasting one to three years, and more communal farmers report that any drought is serious for their survival (Table 5.1).

Table 5.1   Number of years of anticipated drought survival

Farmers responding	*How many years of drought can you cope with?* *Percent of respondents* *(Number of respondents)*				
	0	*1*	*2*	*3*	*4*
Commercial farmers	4	40	32	16	8
(n = 25)	(1)	(10)	(8)	(4)	(2)
Communal farmers	37	34	23	6	0
(n = 35)	(13)	(12)	(8)	(2)	(0)
Total respondents	23	37	27	10	3
(N = 60)	(14)	(22)	(16)	(6)	(2)

Coping strategies (in the short term) and adaptive strategies (in the long term) to manage drought vary, depending upon factors such as culturally determined goals, and the amount and type of resources available to farmers. These resources may be natural or social, as well as economic. Drought coping strategies, more particularly among commercial farmers in the western part of the North-West Province, include maintaining a lower than optimal herd size thus ensuring adequate feed, decreasing herd size, buying feed, or obtaining grazing rights in additional pastureland (communal farmers) (Figure 5.2).

The preferred coping strategy among commercial farmers who feel the need to alleviate drought-related stresses on their herds is to reduce herd size by selling animals. Over twice as many commercial farmers than communal farmers report that they sell animals as their primary strategy in the face of drought. Communal farmers, by comparison, report they are more constrained in their coping strategies, and tend to either buy fodder or just wait until the drought ends. Communal farmers typically report selling animals only when they need household money or in extreme cases drought. When they sell animals in severe droughts, they usually sell only enough animals to get money for maintaining the rest of their herd. Communal farmers therefore usually have strategies that include keeping as many animals as possible.

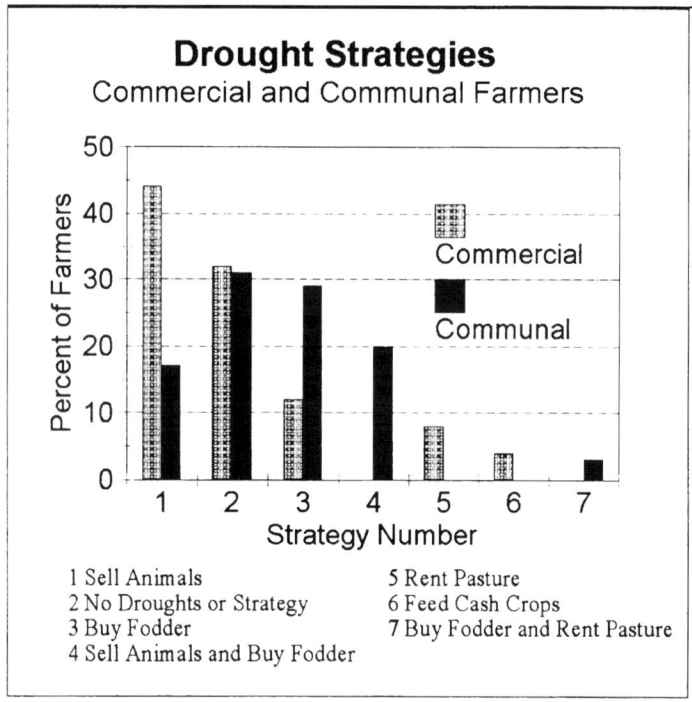

Figure 5.2 Drought coping strategies

No farmers reported keeping cattle species that are more resistant to heat and drought as a strategy to minimize effects of drought. Within both groups of farmers, animal herds of mixed-breeds predominated over single-breed herds. Although a few commercial farmers specialized in single-breed herds, the practice is virtually nonexistent among communal farmers. The practice of having mixed-breed herds, although not reported as an intentional strategy, may be beneficial for farmers in years with ample rainfall as well as in years of drought. In good years with abundant rainfall, less drought-tolerant animal species tend to gain weight faster and reproduce at a higher rate than more drought-tolerant species. Since livestock are usually sold by weight, animals that gain weight and have high reproductive rates in good years benefit farmers in two ways: first, they provide more money per head and second, they contribute to a larger herd size. In a mixed-breed herd, a sufficient number of hardy animals usually survive under the most stressful drought conditions, thus preventing complete herd and financial loss. Inadvertently, the mixed-breed herd strategy can provide a hedge against a total economic disaster.

Within the commercial regions, farms have typically been in the family for three or four generations. With recent harder economic times and decreasing farm profits, there seems to be a growing tendency for farmers to absorb neighboring farms in order to be able to increase herd sizes, and consequently, profit margins. Other farmers diversify by providing products for specific markets such as specialized cattle breeds, producing game animals, operating hunting lodges, or selectively breeding foreign animals. In general, these commercial farms are run for profit, and support a small nuclear family and permanent workers.

A number of factors can be used to explain the reasons for different drought strategies utilized by commercial and communal farmers. The first factor, often given as an overly simplistic reason to explain group adaptive strategies, is that of differing cultural goals. On the one hand, the South African Department of Agriculture has tended to view commercial farmers as successful animal producers who make optimal use of their available resources. Communal farmers, on the other hand, have been viewed by some to over-graze pastures, degrade the environment, and practice methods of inefficient animal production. These ideas are culturally western, based on assumptions that communal farmers tend to keep animals as a measure of wealth and for social status. This research, however, shows reality to be somewhere between these stereotypical views. Communal farmers actually keep animals as a kind of "walking bank account." In good years animals reproduce giving the farmer an average 25% increase in the "bank account," which is vastly higher than selling the animals, depositing the money in the bank and reaping a mere 6% from traditional bank interest rates. Communal farmers tend to dispose of an animal only when they need currency or when the animal is used for household consumption or some other special purpose.

Regardless of coping strategy or amount of available resources, prior knowledge of approaching drought is potentially advantageous for all farmers. For example, with advance knowledge of impending drought, farmers can purchase additional fodder, make advance arrangements for grazing on additional pastureland, or sell animals while market prices are still high and animals in good condition. The role of improved and accessible seasonal climate forecasts thus has an important role to play for better farm management in the southern Kalahari area.

Use of Forecasts in the Livestock Sector

In addition to providing more strategies for drought mitigation, advanced and timely drought information potentially helps farmers develop better drought coping strategies. Nonetheless, this potential is not widely recognized by livestock farmers in this area. Combined as a single group, 60% of farmers in the western part of the North-West Province perceive forecasts being not at all or only somewhat valuable, while 40% perceive forecasts as being valuable or very valuable (Table 5.2). Commercial and communal farmers have, however, different perceptions of the value of forecast information. Disaggregated data show that, 80% of commercial farmers reported forecasts as being not at all valuable or only somewhat valuable, and 54% of communal farmers reported forecasts as being valuable to very valuable. It would seem from this initial assessment that communal farmers value forecasts more highly than to commercial farmers.

Farmers were also asked how they perceived forecast accuracy and if they made, or would make, herd management decisions based on climatic forecast information (Table 5.3). Responses show that 23% of all farmers believe that forecasts are accurate and only 29% said they made or would make management decisions for the 1999-2000 growing season based on climatic forecasts. There is not a high correlation between those farmers who made decisions based on forecasts (29%) and those who say they believe forecasts are accurate (10%).

Table 5.2 Perceived value of climate forecasts

Farmers responding	Value of climatic forecasts Percent of respondents (Number of respondents)				
	Not at all	*Somewhat*	*Valuable*	*Very*	*Total*
Commercial	24 (6)	56 (14)	4 (1)	16 (4)	100 (25)
Communal	31 (11)	14 (5)	37 (13)	17 (6)	100 (35)
Total respondents (N=60)	28 (17)	32 (19)	23 (14)	17 (10)	100 (60)

Table 5.3 Perceptions of forecast accuracy and use in decision-making: All farmers

Decisions made because of forecasts in 1999?	Accuracy of forecasts Percent of respondents (Number of respondents)		
	Not Accurate	*Somewhat Accurate*	*Are Accurate*
No: 71% (n = 43)	33 (20)	25 (15)	13 (8)
Yes: 29% (n = 17)	2 (1)	17 (10)	10 (6)
Total respondents (N = 60)	35 (21)	42 (25)	23 (14)

Forecasts and Drought Impacts: Some Surprises

The El Niño related drought of 1991/92 is called the worst drought of the century in southern Africa by many researchers. The people in southern Africa faced severe food shortages due to drought-related crop failures and threatened the lives of up to 80 million people (Glantz, 1996). The international community and the South African Weather Service did not realize the extent of the possible drought in the 1991/1992 growing season. Concern about potential drought effects was only heightened during the early summer of 1991. Drought information was also not widely distributed to farmers and food-security government officials in South Africa until the summer rains were already overdue (Glantz, 1996; Glantz, Betsill, and Crandall, 1997).

Despite the considerable impacts of ENSO-related drought in the early 1990s, livestock farmers in the western part of the North-West Province overwhelmingly (82%) felt that the El Niño event of 1991-92 had no effect on local agricultural conditions (Table 5.4). Only 16% of the commercial farmers and 17% of the communal farmers reported a severe or very severe effect on the growing season. One explanation for the variation between group reporting is that even minor weather changes could have a larger impact on the more marginalized communal farmers. Another and probably more credible explanation is that group differences could be due to the highly spatially variable nature of rainfall in the area.

When asked if the 1991/92 season was normal (Table 5.5), 73% of the farmers responded that they had a normal year (consistent with the previous question), whereas only 27% stated that the year was not normal. Only about 2% of the farmers thought 1991/92 El Niño had a moderate effect on the weather, 8% thought it had a severe effect, and 8% thought it had a very severe effect, see Table 5.4. Although there are undoubtedly many possible explanations for these findings, two are perhaps the most obvious: the first explanation is that this drought was the first one to be widely publicized as being associated with El Niño, and then, only after the fact. The other explanation is that this drought, while having a disastrous effect on crop farmers in most of southern Africa, had little impact on livestock owners in the western part of the North-West Province. This could be attributed to either a spatial component of the drought or the fact that livestock are much less susceptible to effects of drought than crops, particularly if farmers have access to supplemental feed.

Table 5.4   Farmers' perception of the influence of El Niño on the 1991/92 growing season

Farmers responding	Severity of the 1991-92 EL Niño Percent of respondents (Number of respondents)			
	No effect or n/a	*Moderate*	*Severe*	*Very severe*
Commercial farmers	80	4	12	4
(n = 25)	(20)	(1)	(3)	(1)
Communal farmers	83	0	6	11
(n = 35)	(29)	(0)	(2)	(4)
Total respondents	82	2	8	8
(N = 60)	(49)	(1)	(5)	(5)

Table 5.5   Farmers' perception of season normality for 1991/92

Farmers responding	Was the 1991/92 season normal? Percent of respondents (Number of respondents)	
	Yes	*No*
Commercial farmers	80	20
(n = 25)	(20)	(5)
Communal farmers	69	31
(n = 35)	(24)	(11)
Total respondents	73	27
(N = 60)	(44)	(16)

The 1997/98 Forecasts

In contrast to the drought of 1991/92, even farmers not normally receiving climate forecast information got advance warnings about the anticipated

1997/98 El Niño. The drought that was expected to result from this El Niño event was forecast to be much more severe than any in recent history. This was because the sea surface temperature off the coast of Peru, the primary indicator of El Niño's intensity, rose higher and faster than at any previously recorded time. Forecasters reasoned that the world-wide droughts associated with El Niño would be comparably severe. Consequently, dire warnings were spread in El Niño-associated drought regions, including South Africa. Even farmers living in the most remote villages, who had never heard of or understood the term El Niño, were informed about the upcoming drought (Morobe, 1999). Across South Africa, many farmers undertook preparations in anticipation of the most severe of droughts. This preparation was largely unwarranted, however, because the yearly rainfall was sufficient and resulted in a "typical" growing season.

Some farmers, however, believed the forecasts and took measures to alleviate the approaching drought (Table 5.6). Twelve percent of the commercial farmers and 9% of the communal farmers reported having sustained losses because they prepared for drought. Commercial farmers reported decreasing herd size by up to 30% in order to prepare for a severe drought. Communal farmers, under pressure from government officials, also reduced their animal numbers in preparation for the El Niño-caused drought.

Table 5.6 Losses due to drought in 1997/98

Farmers responding	Losses because of 1997-98 forecast? Percent of respondents (Number of respondents)	
	Yes	No
Commercial farmers	12	88
(n = 25)	(3)	(22)
Communal farmers	9	91
(n = 35)	(3)	(32)
Total respondents	10	90
(N = 60)	(6)	(54)

All of the commercial farmers and 71% of the communal farmers interviewed thought El Niño had "no effect", and that the weather was normal or wetter than normal. Twenty six percent (Table 5.7) of the communal farmers thought the 1997/98 season had been a severe to very severe drought.

Table 5.7 Perceived severity of the 1997/98 El Niño-related drought

Severity	*Percent of respondents (Number of respondents)*	
	Commercial	*Communal*
Severe, very severe	0	26
	(0)	(9)
Moderate, hotter, drier	0	3
	(0)	(1)
No effect, wetter	100	71
	(25)	(25)
Total respondents	100	100
(N = 60)	(25)	(35)

Perceptions of Forecast Information

Despite the value and use of forecasts there are still a number of obstacles and constraints to their utility. One of the most frequently cited problems is that of dissemination and communication of the message. Weather forecasters can be prevented from conveying their information in a variety of ways. Obstacles include: language barriers (not discussed here, but significant to this study), differing types of media, access to that media, reliability, and accuracy of forecast information. Issues such as appropriate scales of the forecasts, timing of forecasts and the "confidence" of the message are added problems.

Commercial livestock farmers tend to want long-term weather forecast information but do not consider the information accurate enough to use as a guide for making changes in their farming practices. Regional variation exists in commercial farmers' desire for forecast dissemination. For example, television and radio coverage is not reliable throughout the Vryburg One region, especially in the extreme north and west part of the district that borders Botswana and Namibia. In this district people tend to receive climatic forecast information in published periodicals such as newspapers and the *Farmers Weekly* (a popular agricultural magazine). Farmers in the Vryburg Two region rely more on information from the Department of Agriculture's Extension Officers but are as skeptical of climatic forecasts as their Vryburg peers. Commercial farmers in both groups see long-term forecasts as not being available during much of the year.

Communal farmers, moreover, tend to rely heavily on the Extension Officers for forecast information. While somewhat skeptical of forecast information they tend to utilize forecasts more than commercial farmers even though the information is less readily available in the communal areas. They would like to see climate forecast information made more widely available with more frequent updates. They list a number of accessibility problems. For example, information is often not provided in their native language, they lack of access to newspapers or magazines, and information is not always available via word of mouth sources, such as village meetings or from extension officers.

It was found that virtually all farmers want climatic forecast information. However, commercial farmers overwhelmingly want this information as supplemental to their own, usually well-established, method of adjusting herd size for optimum calf production. This is in contrast to communal farmers, who state that they wish to have climatic forecast information as a primary guideline for herd management. This was found to be the case even in light of losses incurred because of the drought forecast and warnings issued by the South African Weather Service for the 1997/98 El Niño event.

Conclusions

Droughts and dry spells are regular features of the climate of South Africa. Farmers living in semi-arid and arid environments have therefore developed a number of strategies to help them cope and manage droughts. Improved forecasts, particularly long-term seasonal forecasts and those that indicate potential El Niño-type events, are one of the inputs that farmers can use to help in their management strategies. Examination of the use and uptake of seasonal forecasts by communal and commercial farmers in the North-West province in South Africa show that while forecasts may offer potential benefits, there are a number of limitations that need to be addressed before they will be more useful.

The contexts in which farmers find themselves greatly influence their ability to use and apply forecasts. Commercial farmers, on the one hand, still have the ability, by way of their access to resources, to manage their risks in a more risk-averse manner, despite changes in farming assistance and land management in South Africa. Twice as many commercial as communal farmers therefore reported that they do not need to make many major changes during drought periods. Most indicate that they can cope with a drought of 1-3 years duration. Communal farmers, on the other hand, have often been denied similar access to resources and therefore have usually managed risks by trying to maintain herd sizes during dry years, expanding them when conditions improve. Communal farmers therefore report that any drought is serious for their survival and they tend to keep animals to "see them through the drought".

The science of forecasting, particularly seasonal forecasting has greatly improved over recent years (Mason et al., 1996; Unganai, 1996; Mason, 1997; Klopper, 1999). Potential users, however, still face obstacles in their ability to use El Niño and other long-term forecast information for their decision-making processes. Such obstacles include accuracy of forecasts, delays in the timing of forecasts, and concerns about the value of regional-scale forecasts for local-level decision making. Most farmers consider forecasts, in their current form, as not useful. The availability and systematic distribution of forecasts, including the media through which forecasts are available, the language in which forecasts are disseminated, and problems interpreting the meaning of forecasts may be possible explanations for these perceptions. One third of all farmers, however, made some management decisions based on the 1999/2000 growing season. If forecasts are going to be more widely used in the southern Kalahari region of South Africa, serious considerations need to be made about improved understanding of the socio-economic context within which farmers operate and make decisions, and the communication and dissemination of climate forecast information.

Acknowledgments

This chapter is dedicated to Jim Ellis, a key principal investigator from the Natural Resources Ecology Laboratory, Colorado State University, who gave valuable assistance and insights to the project and who tragically lost his life in a skiing accident in March 2002.

Funding for this thesis research was provided by the National Oceanic and Atmospheric Administration (NOAA) Grant DOC-NOAA NA86GP0347 AMEND #1.

Notes

1 Many types of fast-growing annual grasses, locally known as "two week grasses", are widespread in South African rangelands and provide ample nourishment for animals while slower-growing perennial grasses develop.

2 Under drought conditions, animals often eat poisonous plants and poisoning episodes have been reported by farmers in all five districts studied. For example, in the drought of 1969, over 600,000 sheep in South Africa died of poisoning from plants they normally avoid eating. During the same drought, a farmer in the Kudumane district supplemented the diet of about 40 cows with dune bush, a poisonous plant species that cows usually avoid. This resulted in death of nearly half of his animals (Kellerman, Coetzer, and Naude, 1988). Thus, El Niño events can influence the species composition of forage as well as forage availability.

3 The commercial farmer reporting poor water quality is located in the extreme northwest part of the Vryburg 1 district. This farmer reported having water that he considered unfit for both human and animal consumption due to its high salt content. The three communal farmers reported water quality to be bad because of contamination from animal waste.

References

Acocks, J.P.H. (1975), *Veld Types of South Africa* (3rd edition), Botanical Research Institute, Pretoria.

Bromley, D.W. (1995), 'South Africa - Where Land Reform Meets Land Restitution', *Land Use Policy*. special issue, vol. 12, no. 2.

Bruwer, J.J. (1989), 'Drought Policy in the Republic of South Africa, Part 1', *Drought Network News*, vol. 1, No.3, pp. 14-16.

Changnon, S. A. (ed.) (2000), *El Niño 1997 – 1998 The Climate Event of the Century*, Oxford University Press, New York.

Cowling, R. M., Richardson, D.M., and Pierce S.M. (1997), *Vegetation of Southern Africa*. Cambridge University Press, Cambridge.

Freeman, C. (1984), 'Drought and Agricultural Decline in Bophuthatswana', in South African Research Services, (eds), *South African Review II*, Ravan Press, Johannesburg, pp. 248-289.

Glantz, M.H. (1996), *Currents of Change: El Niño's Impact on Climate and Society*, Cambridge University Press, Cambridge.

Glantz, M., Betsill, M. and Crandall. K. (1997), *Food Security in Southern Africa: Assessing the Use and Value of ENSO Information*, NOAA Project, National Center for Atmospheric Research, Boulder, CO.

Harsch, E. (1992), 'Drought Devastates Southern Africa', *Africa Recovery*, April 1992.

Hulley, A.G. (1980), 'Drought Modelling in South Africa', M.Sc Thesis, University of Natal, Durban.

Kellerman, T.S., Coetzer, J.A.W., and Naude, T.W. (1988), *Plant Poisonings and Mycotoxicoses of Livestock in Southern Africa*, Oxford University Press, Cape Town.

Klopper, E. (1999), 'The Use of Seasonal Forecasts in South Africa during the 1997/1998 Rainfall Season', *Water SA*, vol. 25, no. 3, pp. 311-316.

Laing, M.V. (1992), 'Drought Update 1991- 1992: South Africa', *Drought Network News*, vol. 4, pp. 15-16.

Leppan, H.D. and Bosman, G.J. (1923), *Field Crops in South Africa*, Central News Agency, Ltd., Johannesburg.

Lindesay, J. (1998), 'Present Climates of Southern Africa', in J.E. Hobbs, J.A. Lindesay, and H.A. Bridgeman, (eds), *Climate of the Southern Continents, Present, Past and Future*, Wiley, New York, pp. 5-62.

Mans, Dirk. (2000), South African Department of Agriculture, Field Extension Officer, personal communication.

Mason, S. (1997), 'Review of Recent Developments in Seasonal Forecasting of Rainfall', *Water SA*, vol. 23, pp. 57-62.

Mason, S.J., Joubert, A.M., Cosijn, C. and Crimp, S.J. (1996), 'Review of Seasonal Forecasting Techniques and their Applicability to Southern Africa', *Water SA*, vol. 22, no. 3, pp. 203-209.

Mason, S.J. and Jury, M.R. (1997), 'Climatic Change and Interannual Variability over Southern Africa: A Reflection on Underlying Processes', *Progress in Physical Geography*, vol. 21, pp. 24-50.

Morobe, W. (1999), Information Officer, South African Department of Agriculture North - West Province, Western Region, Vryburg, South Africa, personal communication.

National Center for Atmospheric Research (NCAR) (1985), 'African Drought: Monitoring and Prediction', Workshop Report, NCAR, Boulder, CO.

Patt, A. (2001), 'Understanding Uncertainty: Forecasting Seasonal Climate for Farmers in Zimbabwe', *Risk Decision and Policy*, vol. 6, no. 2, pp. 1-15.

Phillips, J.G., Makaudze, E. and Unganai, L. (2001), 'Current and Potential Use of Climate Forecasts for Resource-poor Farmers in Zimbabwe', in American Society of Agronomy Special Publication #63. *Impacts of El Niño and Climate Variability on Agriculture*, pp. 87-100.

Roberts, B.R. and Fourie, J.H. (1989), *Common Grasses of the Northern Cape*, Northern Cape Livestock Co-operative Limited, Kimberley.

Rwelamira, J.K. (1997), 'Background Paper on the Economic and Financial Implication of Drought to the State and to the Economy of South Africa', Ministry of Agriculture, Pretoria.

Simbi, T. (1998), *Agricultural Policy in South Africa - A Discussion Document*, Ministry for Agriculture and Land Affairs, Pretoria.

Tyson, P.D. (1986), *Climatic Change and Variability in Southern Africa*, Oxford University Press, Cape Town.

Unganai, L. (1996), 'Recent Advances in Seasonal Forecasting in Southern Africa', *Drought Network News*, June 1996.

Van Zyl, J., Biswanger, H.P and Kirsten, J.F. (eds) (1996), *Agricultural Land Reform in South Africa: Policies, Markets and Mechanisms*, Oxford University Press, Cape Town.

Vogel, C. (1993), 'A Prayer for Rain in Bophuthatswana', *Indicator South Africa*, vol. 10, no. 4, pp. 52-54.

Vogel, C. (1994), 'South Africa', in M.H. Glantz. (ed.), *Drought Follows the Plow*, Cambridge University Press, Cambridge, pp. 151-170.

Vogel, C. (2000), 'Usable Science: An Assessment of Long-term Seasonal Forecasts amongst Farmers in Rural Areas in South Africa', *The South African Geographical Journal*, vol. 82, pp. 107-116.

Vryburg Development Center (1999), *Bophimira Agriculture Tours*, Edson Clyde Press, Vryburg.

Wilhite, D. (2000), 'Drought as a Natural Hazard: Concepts and Definitions' In D. Wilhite, (ed.), *Drought: A Global Assessment*, vol. 1, Routledge, London, pp. 1-18.

6 Climate Forecasts in Swaziland: Perspectives from Agribusiness

LOUISE BOHN

Introduction

Seasonal forecasts are viewed as a management tool that can assist in improving resilience to rainfall variability and climate in general. Interest in forecast use has grown in southern Africa as the capacity to produce climate forecasts has improved (Mason, 1998; Barnston, Glantz, and He, 1999). However, compared to developed countries, where research into climate forecast use has been extensive (Washington and Downing, 1999), there is still a large amount of material to be collected and interpreted in southern Africa. This is particularly true in terms of case studies on the use of climate information for specific sectors. One potential user group that has been largely ignored in research on forecast use is agribusiness.

Agribusinesses, which represent part of the commercial agricultural sector, are likely to have access to both forecasts and resources for scientific interpretation of the information. On the one hand, it can be argued that these commercial users have the greatest capacity to adapt production to climate variability because of financial and technological resources. On the other hand, agribusinesses may not have the flexibility to respond to individual forecasts because they have a more rigid infrastructure and strict budgets that are tied to planning. Thus, climate forecasts could have great benefits for agribusinesses, but at the same time agribusinesses may face disproportionately large constraints to forecast use.

In this chapter, the use and uptake of information from seasonal climate forecasts is examined for six agribusinesses and one water-resource management company in Swaziland. The agribusinesses include three sugar cane estates, a forestry plantation, a citrus-fruit grower, and a cattle and game ranch.[1] The results presented here form part of a wider assessment of the value of forecast information in southern and eastern Africa (Bohn, 2000). The potential use of climate information is discussed, and constraints to the use and applicability of forecasts are highlighted. Swaziland serves as an ideal case study, as it covers a relatively small geographical area but has a varied topography and climate. Except for the forestry company, the agribusinesses studied are all found in the Lowveld region in the east of the country. The wet season in this region runs from October through March, and all sites receive between 600 and 750mm average annual rainfall. Forestry activities are usually located in the Middleveld region, which receives an annual average rainfall of 850 mm. Water resources are scarce enough in this region for their efficient use to be of concern.

Assessing the Use of Climate Information

Assessing how climate information is used typically involves one of two methods: either addressing the question of forecast use after the season or event has occurred, or questioning users about hypothetical, "perfect" forecasts (Mazzocco et al., 1992; Kite-Powell and Solow, 1994; Barrett, 1998). However, hindsight can cloud judgment and perceptions when assessing forecast use after the season or event has occurred, and hypothetical forecasts can be unrealistic, assuming levels of certainty that are not available given current scientific skill. In fact, the predictability of the region's climate has decreased in recent years (Mason, 1997) making "perfect" forecasts even more unrealistic. A third and alternative method involves real-time evaluations of climate forecast use. This method can provide realistic assessments of climate forecasts that take into account the actual context for forecast use and uptake.

To develop an understanding of the potential for forecast use in agribusinesses, seasonal timelines were used as the focal point during interviews. The objective of these seasonal timelines, sometimes referred to as seasonal calendars (see Chambers, 1992), was to determine the use and value of forecasts in decision-makers' strategies. Timelines are a particularly effective method for extracting 'added value information' from users about forecasts (Bohn, 2000).

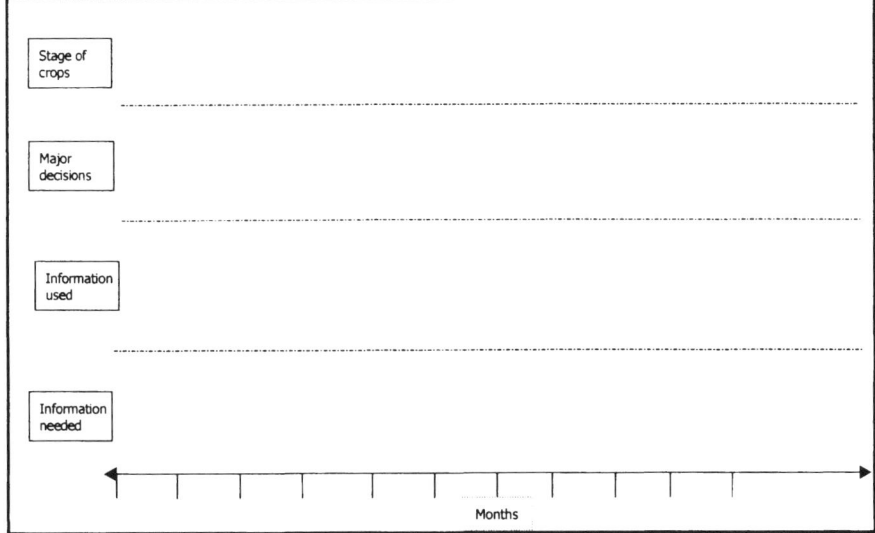

Figure 6.1   Blank timeline presented to interviewees

A blank timeline was presented to representatives from the agribusinesses addressed in this study (Figure 6.1). As a schematic, these timelines have the advantage of being relatively simple for users to construct and interpret. They can be used to extract extensive and detailed information that can help to identify and assess climate-related decisions. Timelines can also be used to address some of the "how" and "why" questions related to forecast use. For example, how do users access forecast information? And why are they using or not using the forecasts?

After timelines had been constructed and background information had been collected, a real-time study of forecast use was undertaken at the start of the 1998/99 rainy season in Swaziland in order to assess constraints to forecast use. Users or potential users were asked for their opinion on the October-November-December (OND) and January-February-March (JFM) 1998/99 forecasts in the week following the October 1998 Southern Africa Regional Climate Outlook Forum (SARCOF) in Harare, Zimbabwe. The responses thus reflect opinions at the time of the event in response to the current state of forecast science. The results can be used to assess the operational usefulness of climate forecasts in their current form to agribusinesses.

Seasonal Forecasts: The Potential

The timelines shown in Figure 6.2 and Figure 6.3 depict examples of the operations, decisions, climate information used, and climate information needed over an agricultural year for two agribusinesses. For forestry (Figure 6.2), the main operations throughout the year relate to the growth and planting of seedlings as well as maintenance of established forests. In addition, fire protection has to be undertaken during drier months. The main decisions relating to these operations are financial, with budget decisions influencing the following season's agricultural program. For fire protection, detailed information about daily temperature, wind and humidity conditions is required. Information about rainfall and wind is also needed for the other forestry operations. Seasonal forecasts have the potential to be used for planning purposes, specifically the timing of planting over the rainy season. Given the long term nature of forestry (which in contrast to the other agribusinesses operates on decadal scales), forecasts that indicate changes in longer term climate trends would also be useful.

The water management company's operations involve the maintenance of both canals and water supply infrastructure, the refill of the storage reservoir during the rainy season, and operations under peak demand for water. Although decisions must be made about expenditures, the primary concern is pumping strategies. This is determined by both the demand for water and water availability; factors that are linked to climate conditions. During the dry season it may be necessary to set up restrictions on water use because of limited supply and increased demand. In order to manage water resources efficiently, the company uses both historical information about the river catchment it extracts water from as well as seasonal forecasts of temperature, river flow and rainfall. Although the water management company already uses rainfall forecasts such as those issued by SARCOF, these forecasts rarely have a big influence on decisions as they do not provide the right information. The company would like more frequent forecasts that are available throughout the year, with improved accuracy and more details towards the end of the wet season.

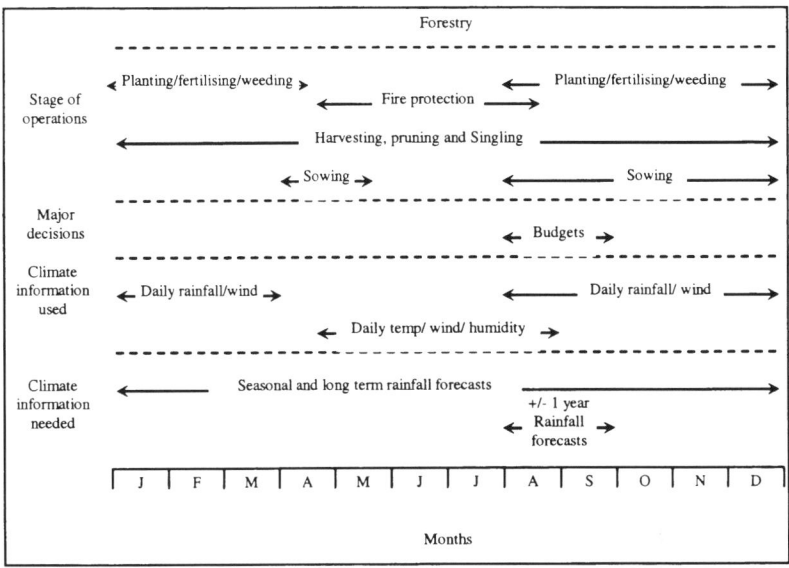

Figure 6.2   Example of a completed seasonal timeline for a forestry company

The operations affected by climate variability are summarized for each type of business in Table 6.1. In general, there is potential to incorporate seasonal forecasts into the planning process in all of the agribusinesses studies. Forestry requires seasonal and long-term forecasts for the whole year as well as more detailed forecasts at the very start of the wet season. Sugar cane producers could use forecasts both for long-term planning such as water management issues and yield estimates, as well as for decisions such as when to harvest. Forecasts could also feed into ranching decisions where the amount of rainfall influences the vegetation available which in turn influences the ranches cattle stocking rates. Even within a small area such as the Swaziland Lowveld, where the three major sugar producers grow sugar cane, there were noticeable differences in perceived requirements. All three sugar cane producers wanted different information on different time scales.

Despite the potential outlined above, there are, however, constraints to forecast use, which are discussed in the next section. Constraints are defined here as barriers to effective forecast use. They can occur at any stage in the forecast process, from forecast production to end user application. Constraints include essentially any factor that limits a user's ability to respond to a forecast.

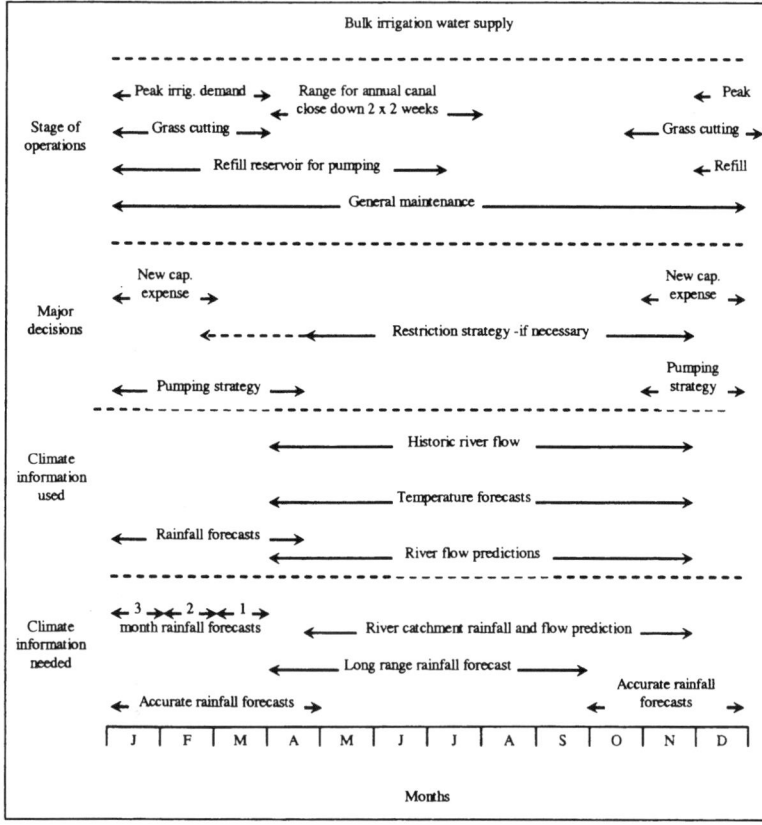

Figure 6.3   Example of a completed seasonal timeline for a water management company

Seasonal Forecasts: The Constraints

While the potential benefits of seasonal forecasts clearly emerge from the timelines, when questioned in "real time" about the SARCOF forecasts, respondents revealed a number of constraints related to forecast use. The constraints discussed here form what may be termed the "constraints gap." This is the gap that exists between the notional value of the forecast information and the true value of the forecast to the user. In order for the notional and true values to converge, it is necessary to first identify the constraints. The constraints that emerged from the interviews include forecast timing, scale, quality, type of event, and risk and uncertainty. Once these constraints have been addressed, a strategy can be developed to reduce the gap and thus optimize forecast use.

Table 6.1 Agricultural operations and decisions affected by climate information

Sugar Cane	Water Management
Harvesting	Canal maintenance
Burning	Storage scheduling
Ripening	Pumping
Dry-off	Planning restrictions etc
Yield and estimates	
Ripping and land preparation	**Cattle/Game Ranch**
Irrigation scheduling	Vegetation quality and quantity
Water management issues	Vegetation burning
	De-stocking/marketing
Citrus Fruits	
Fruit flowering/set	**Forestry**
Cell division/expansion	Fire risk
Harvesting	Planting seedlings
Spraying	Harvesting
Irrigation scheduling	Sowing

1. Timing of the Forecasts

The timing of the forecasts is crucial if they are to be used effectively by agribusiness. SARCOF forecasts were available in October 1998, yet for most businesses it was during July and August when they would have been most useful. For example, concerns about irrigation are greatest in July and August, when most water is extracted from the reservoir. Some users need to know whether water resources will be limited at the end of the dry season due to late rains, or whether it is safe to use up the irrigation quota over the normal time period. As one respondent remarked; "If we had this [SARCOF] forecast ... in August/September, we would have changed things... It is now ideal planting conditions but we don't actually have a lot of land prepared for planting."

The timing of mid-season forecast updates is seldom appropriate for management changes in agribusiness. Agribusiness involves a chain of activities that go beyond planting and harvesting. Too much rainfall, for example, may hamper milling activities in the sugar industry. Agribusiness also makes key production decisions at the start of the season, and it is too late to change those decisions based on an updated forecast. The potential to alter production decisions depends on the flexibility of operations. Yet flexibility is often limited, as budgets have usually been finalized by the time the forecasts are released. Regarding the October forecasts, one manager pointed out that "in terms of our budgeting year, this is a little bit late already ... we really need to have the best forecast we possibly can at the start of September."

The research outlined here suggests that the late timing of the forecasts usually compromises effective use of the forecasts. A balance therefore needs to be struck between the practical application of a forecast and the scientific skill associated with that forecast. Many studies have examined the compromises users are willing to make between forecast timing and quality (Easterling and Mjelde, 1987) but this has not been assessed in southern Africa.

2. Spatial scale of the forecasts

Spatial scale is a critical issue, especially for a small country such as Swaziland. SARCOF forecasts sometimes correspond to large areas such that all of Swaziland is usually included in one forecast region. There are, however, at least three distinct climatic zones within the country due to the varied topography. At other times the boundary between two different forecast regions runs through Swaziland, making it difficult to know which one is applicable to which part of the country.

Consequently, there is a need to identify the scale at which a forecast achieves optimal benefits. If the spatial scale is too coarse, then it may be of little use to specific agribusinesses. Given local knowledge of Swaziland's climate, users do not feel confident that the forecasts can accurately capture regional climatic differences. High-resolution forecasts that would be preferred by individual users require considerable computing data and resources.

3. Reliability of Forecasts

Spatial and temporal limitations of forecasts are further compounded by problems related to the perceived reliability of the forecasts. The lack of belief in the forecasts emerged as a crucial constraint for agribusiness. From a user's perspective, little is known about the quality of the forecasts. In the agribusinesses interviewed for this study, some indication of the reliability of the forecast was often requested. As one manager put it, "we would like to see how good [the forecast] actually is, and we need more information about the skill that is involved. We would like to see what actually happened, like for last year for example." At present, users' evaluations of forecast quality largely depend on how they perceive the performance of past forecasts. This, in turn, is often influenced by the media's portrayal of a forecast and the outcomes.

A forecast that was perceived as inaccurate, as was the case with the 1997/98 El Niño forecast, can influence future uptake and use of forecasts. Many users were keen to know why the 1997/98 forecast did not reflect actual conditions, i.e., why the event had not been as extreme as they had been led to believe. The potential use of a forecast may arguably be improved if reasons behind a poor forecast–such as unexpected changes in Indian Ocean sea surface temperatures–are explained (see Chapter 1, this volume). As one respondent expressed, "we remember what happened, and if the last El Niño had come right, then I think that people in Swaziland would make serious decisions" based on the forecasts. In other words, there is a general understanding among agribusinesses that "weather prediction has to prove itself."

4. The 'Risk Environment'

The agribusinesses interviewed for this study generally tend to be risk averse when considering forecast applications. The degree of certainty that can be assigned to seasonal forecasts is not always sufficient to merit acting upon them. Agribusinesses dislike taking risks, and the tercile system used in disseminating SARCOF forecasts is perceived as highly uncertain.[2] Managers working with operations using irrigated agriculture are often less likely to take risks based on a forecast, as they already have some insurance against drought impacts. They require a particularly high forecast probability before acting. As one manager acknowledged, "a forecast with a 60% probability we would view with interest, but it would realistically need a probability of 80% to influence us."

Response options and the consequences of actions influence risk-taking behavior. In terms of acting upon forecasts, there are different risks for different sectors. Forestry, for example, can risk acting on climate forecasts because if they are wrong, the losses are not that great and normal planting can usually resume. If the forecasts are correct, however, then planting may be optimized. They are less risk averse in this situation because the losses would be minimal. Agribusinesses relying on irrigated agriculture can respond incrementally and monitor progress as the season develops, rather than committing to long-term actions at seasonal timescales.

5. Type of Event

The degree of response to forecasts is also dependent on the type of event that is of concern to each agribusiness. For the sugar companies, it is rain rather than drought that can cause the most operational problems. Too much rain can cause mechanical problems for harvesting and milling, which can have a severe impact on yield. Forecasts of both the probability of above-normal and below-normal rainfall are therefore equally useful for these producers (see Chapter 7, this volume).

Water is generally considered a scarce commodity in Swaziland, thus the overriding concern is directed towards below-normal rainfall forecasts.

Interestingly, far more caution is applied when reacting to projections of above-normal rainfall. It is considered safer to conserve water regardless of the forecast, rather than be faced with an unexpected shortage. Conserving water during a year with above-normal rainfall is less costly than increasing water consumption during a year with below-normal rainfall.

Conclusions

There is often an underlying assumption that seasonal forecasts are useful, especially to large agribusiness. The findings discussed here, however, question whether this sector, which has perhaps the best access to these forecasts, actually makes full use of them or whether they have the capacity or incentives to do so. For irrigated farms, users are often more concerned with the impact of climate on mechanical operations, as opposed to the efficient use of water resources. Given fixed infrastructure, there is little flexibility in their response. Sugar producers are unlikely, for example, to reduce or increase the area of cane planted for one year as a result of a forecast.

Budgets are also an important consideration; although agribusinesses may have more funds available than smaller subsistence farmers, resources are strictly allocated. Budgets often dictate operations, overriding any decision due to forecasts. Some agribusinesses, for example, may have to reduce plantings one year because of budget constraints, regardless of the seasonal climate forecast.

In the context of Swaziland agribusiness, constraints to forecast use can be summarized as follows:

- The timing of the forecasts is often too late;
- The spatial extent of the forecasts is too coarse;
- Lack of verification is an issue;
- Perception of past performance affects use;
- Probabilities are not high enough to merit action;
- Risk averseness relates to available responses and possible impacts;
- Interpreting probabilities can cause confusion; and
- The event that is forecasted influences forecast usefulness.

Interestingly, few constraints are related to the external socio-economic and political environment in which the agribusinesses operate, unlike the resource-poor farmers described in other chapters in this book. Access to resources, whether financial or agricultural, for example, was not considered to be a major issue. Although none of the companies have a specific policy relating to forecast use, agribusinesses were generally receptive to forecasts, and forecasts were viewed positively, despite their limitations.

The overall position of agribusinesses can be summed up by this response: "If we had a useful forecast, we could act, and we would act." This contrasts with the experience of the smaller-scale user, who may have less access to forecasts and fewer resources to act upon the forecasts. For the small-scale user, for example, economic factors may present greater constraints than those related to the forecasts, and it may be a case of "even if we had a useful forecast, we couldn't act."

This research shows that, for the businesses covered here, even where forecast information is available, it cannot always be used. It is crucial therefore that there is a greater understanding of all the constraints to forecast use. To optimize the value of seasonal climate forecasts, these constraints need to be addressed, and the "constraints gap" reduced.

Notes

1 Many of these agribusinesses receive investments from the Commonwealth Development Corporation in London. This company has a strong interest in the application of forecast information and provided the initial impetus for the research.

2 Communicating forecast uncertainty while still maintaining end user confidence is considered a real challenge (Barrett, 1998). The possibility of poor interpretation of probabilities can lead to a misunderstanding of the risk. For example, a common error is to add together the probabilities of two of the categories. If the forecast was 30% above, 30% near and 40% below a user might add the near and above normal together thus assuming 60% of 'not below.' This would outweigh the correct probability of 40% below normal.

References

Barnston, A.G., Glantz, M.H. and He, Y. (1999), 'Predictive Skill of Statistical and Dynamical Climate Models in SST Forecasts during the 1997-98 El Niño Episode and the 1998 La Niño Onset', *Bulletin of the American Meteorological Society*, vol. 80, pp. 217-243.

Barrett, C.B. (1998), 'The Value of Imperfect ENSO Forecast Information: Discussion', *American Journal of Agricultural Economics*, vol. 80, pp. 1109-1112.

Bohn, L.E. (2000), 'The Use of Climate Information in Commercial Agriculture in Southeast Africa', *Physical Geography*, vol. 21, pp. 57-67.

Chambers, R. (1992), 'Rural Appraisal: Rapid, Relaxed and Participatory', Institute of Development Studies, Discussion Paper 311, University of Sussex.

Easterling, W.E. and Mjelde, J.W. (1987), 'The Importance of Seasonal Climate Prediction Lead Time in Agricultural Decision-making', *Agricultural and Forest Meteorology*, vol. 40, pp. 37-50.

Kite-Powell, H.L. and Solow, A. (1994), 'A Bayesian Approach to Estimating Benefits of Improved Forecasts', *Meteorological Applications*, vol. 1, pp. 351-354.

Mason, S.J. (1997), 'Recent Changes in El Niño Southern Oscillation Events and their Implications for Southern African Climate, *Transactions of the Royal Society of South Africa*, vol. 52, pp. 377-403.

Mason, S.J. (1998), 'Seasonal Forecasting of South African Rainfall using a Non-Linear Discriminant Analysis Model', *International Journal of Climatology*, vol. 18, pp. 147-164.

Mazzocco, M.A., Mjelde, J.W., Sonka, S.T., Lamb, P.J., and Hollinger, S.E. (1992), 'Using Hierarchical Systems Aggregation to Model the Value of Information in Agricultural Systems - An Application for Climate Forecast Information', *Agricultural Systems*, vol. 40, pp. 393-412.

Washington, R. and Downing, T.E. (1999), 'Seasonal Forecasting of African Rainfall: Prediction, Responses and Household Food Security', *The Geographical Journal*, vol. 165, pp. 255-274.

7 Determinants of Forecast Use among Communal Farmers in Zimbabwe

JENNIFER PHILLIPS

Introduction

Agricultural production in Zimbabwe is high relative to most other southern African countries because of the combined output of both the high-input commercial agricultural sector and the low-input smallholder component, each of which contribute roughly half of the national grain output (Masters, 1994). Yet in spite of its reputation as the "breadbasket" of southern Africa, food insecurity is still a regular occurrence in Zimbabwe, particularly in the drier, southern provinces.

Early hopes within the food security community were that seasonal climate forecasts would help alleviate hardship during poor rainfall years through improved farm management. However, a number of questions have arisen concerning who is actually able to benefit from forecasts, and under what conditions (Pfaff, Broad, and Glantz, 1999). If the intended beneficiaries of forecasts are the neediest, one may ask whether the neediest are able to respond to climate information, or indeed, whether they receive the information in the first place.

This chapter focuses on whether and to what extent resources, or "wealth," affect how climate forecasts are used by smallholder farmers in Zimbabwe. Zimbabwe was one of the first countries in Africa to begin officially disseminating seasonal climate forecasts. The strong El Niño event of 1997/98 and the subsequent La Niña in 1998/99 provided an opportunity to observe forecast responses in the farm sector in both below normal and above normal rainfall years. The underlying agroecological conditions are examined in relation to forecast type to determine their influence on forecast-based decisions. Results from household surveys carried out during the 1997/98 and 1998/99 El Niño Southern Oscillation (ENSO) events form the basis of the analysis.

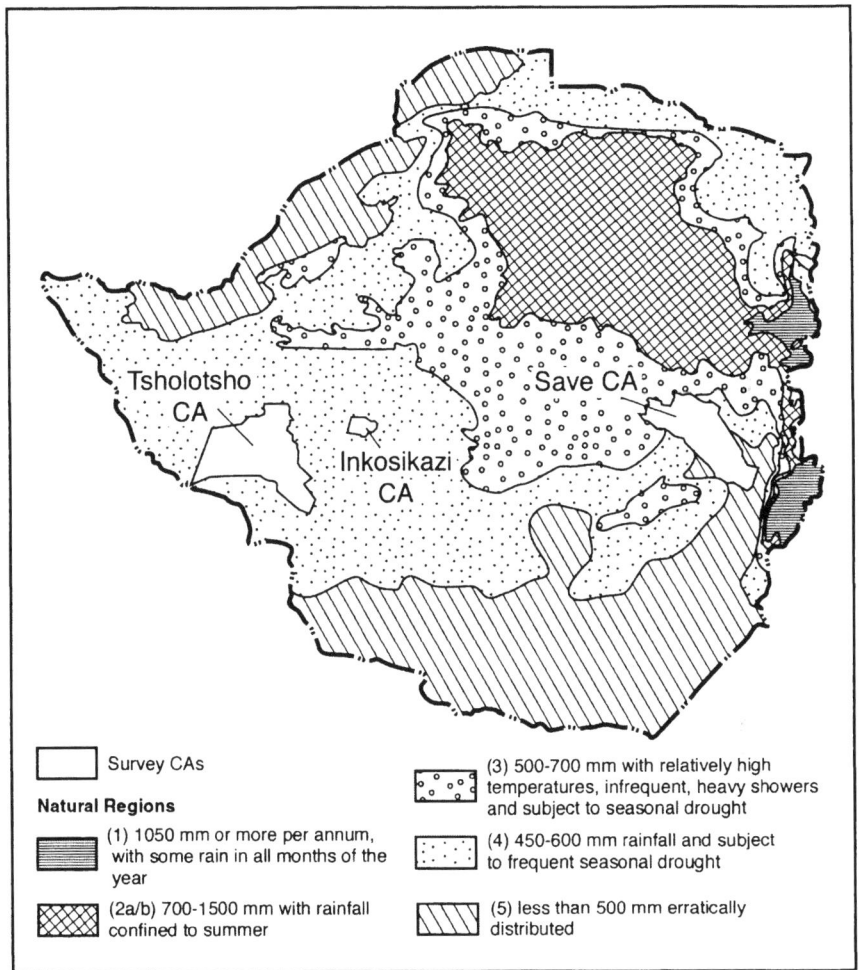

Figure 7.1   Map of Zimbabwe showing survey communal areas (CAs) and Natural Regions

Background

The agricultural sector in Zimbabwe is distinctly divided between a high-input, commercial sector and a low-input, smallholder sector. The commercial farm sector is dominated by farmers of European descent who remained in the country after Zimbabwe gained independence in the late 1970s. Levels of input use and yields of the major grains (maize and irrigated wheat) are comparable with those of the United States. Because it is assumed that commercial farmers have adequate access to seasonal climate forecasts and other information, and that they have the resources–including credit and crop insurance–to be able to take adaptive action if necessary, they have not been the main focus of research. The smallholder sector, on the other hand, has fewer resources at its disposal and is thus the subject of considerable scholarly attention. This chapter will thus focus on smallholder farmers in Zimbabwe, henceforth referred to as "the communal farm sector."

The communal farm sector in Zimbabwe developed on what were called "Tribal Trust Lands" allocated to black farmers during the colonial era, usually in the less productive zones. Communal Areas (CA) are disproportionately located in the lower rainfall zones. Using the classification of Vincent and Thomas (1960), Zimbabwe is divided into five agroecological zones called Natural Regions (NR), based on annual rainfall totals, rainfall reliability, and soil type (Figure 7.1). The categorization of the Natural Regions reflects the inherent productivity of the site, with highest productivity in NRs 1 and 2 (where rainfall ranges from 700 mm per year to over 1500 mm per year and is relatively uniform throughout the season), and lowest productivity in NR 5 (where annual rainfall is less than 500 mm and midseason dry spells are frequent). Ten percent of CAs are located in NRs 1 and 2, 17% in NR 3, and 73% in NRs 4 and 5 (Ministry of Lands, Agriculture and Rural Resettlement, 1990).

Farming systems in the Communal Areas correspond to the Natural Region in which they are located, with maize dominating in NRs 2 and 3, and small grains becoming increasingly important in zones 4 and 5.[1] Although NR 5 is considered to be best suited for cattle grazing, drought-tolerant crops such as pearl millet and sorghum are grown, as well as small amounts of short season maize. Other crops commonly grown in Communal Areas include groundnuts, cotton, and sunflower.

The use of agricultural inputs in the communal farm sector is limited. Fertilizer use is currently extremely low, and tends to be concentrated on maize (Huchu and Sithole, 1994; Scoones, 1996a). This is true for both chemical fertilizers, which are no longer subsidized by the government, and manure.[2] One of the few purchased inputs is hybrid maize seed. Close to one hundred percent of Zimbabwean small farmers plant hybrid maize, which is partly responsible for the growth in production by the communal farm sector since independence (Masters, 1994; Phillips, Makaudze, and Unganai, 2001). Nonetheless, average maize yields are only about one ton per hectare (which is considerably higher than yields of small grains). Poor soil fertility contributes to low yields, as does the late planting date which results from poor access to draft animals. Labor is a further constraint for some households, as family members often leave to seek work in urban areas (Kinsey, Burger, and Gunning, 1998).

The Survey

A survey was carried out among Communal Area farmers in the 1997/98 and 1998/99 cropping seasons to investigate access to climate information and the use of forecasts in farm management. During the 1997/98 rainy season, 225 households were interviewed in October while the cropping season was being planned, and in May after harvesting. Seventy-five households were interviewed in each of NR 3, 4, and 5 in Save Communal Area, which is located in Shona-speaking territory and runs north-south along the decline from the high veldt to the low veldt in the east-central region of Zimbabwe (for additional details, see Phillips, Makaudze, and Unganai, 2001). In 1998/99, the survey sample was doubled to 450 households. The same families in the three NRs in Save CA were revisited, and an additional 75 families were selected in each of Murewa CA (NR 2), Inkosikazi CA (NR 4) and Tsholotsho CA (NR 4) (Figure 7.1). The latter two Communal Areas are in the Ndebele-speaking region to the west, and represent a somewhat distinct cultural and farming group. Enumerators were selected from the local community and participated in a three-day training program in which survey instruments were tested. Households were selected at random using the most recent census, and were distributed across at least three wards per survey site.

Data on household assets, access to climate information, local customs for rainfall prediction, and intentions for the coming season's crop management were gathered. To evaluate deviations from standard management practices, farmers were asked a number of questions regarding "normal" crop management. The analysis presented here stratifies households according to assets, Natural Regions, and forecast year (the 1997/98 El Niño and the 1998/99 La Niña) in an attempt to identify the primary determinants of strategies identified by farmers as appropriate responses to climate forecast information. Gender differences were also assessed. Overall, 29% of the households surveyed were headed by women. The high percentage of female-headed households may reflect both the increasing use of off-farm sources of income, and also the toll taken by HIV/AIDS in Zimbabwe.

Although work by Scoones (1995) has shown that farmers' own perception of wealth may differ from that pre-identified by researchers, factors included in the "assets" variable were selected on the basis of their likely relationship to crop management options. The variable indicating assets is thus defined as the simple sum of the number of draft animals owned and the number of hectares of arable land for the household. Households were categorized into three groups of roughly equal size, reflecting low (1), moderate (2), and high (3) levels of assets, relative to the range of assets found in this population.

Cattle ownership is often considered the most important determinant of wealth among rural communities in Zimbabwe, and is directly related to farm management through timeliness of plowing and planting, and amount of manure applied (Carter and Murwira, 1995). Furthermore, cattle are increasingly used for coping with climate-related crop production risk through sales to supplement income (Kinsey, Burger, and Gunning, 1998). However, as cattle populations have dwindled over the last decade, other factors of production have grown in importance (Scoones, 1995). Land is not formally owned in the Communal Areas, and fields are assigned by the village chief. The number of hectares of arable land is likely to reflect the degree of flexibility in area-planted decisions and serve as a measure of influence in the village. Although Dalton, Masters, and Foster (1997) found that labor is readily substituted for other inputs in agricultural production in the communal farm sector in Zimbabwe, in this survey labor (represented by the number of adults per household) was strongly positively correlated with draft animal ownership and land area cultivated. As a result, labor was not included in the assets variable.

Results

1. Distribution of Assets

As described above, the sum of draft animals and area of arable land was used to create an "assets" variable, which can be considered a measure of opportunity for changing crop management when climate forecasts are available. Assets in NR 3 are greatest (6.76 units per household), followed closely by assets in NRs 2 and 4 (5.43 and 5.08 units per household, respectively). Assets in NR 5 amount to less than half of those in NRs 2 and 4, and roughly one third of those in NR 3.

The composition of assets varies across NR, as shown in Figure 7.2. The number of draft animals owned per household decreases with Natural

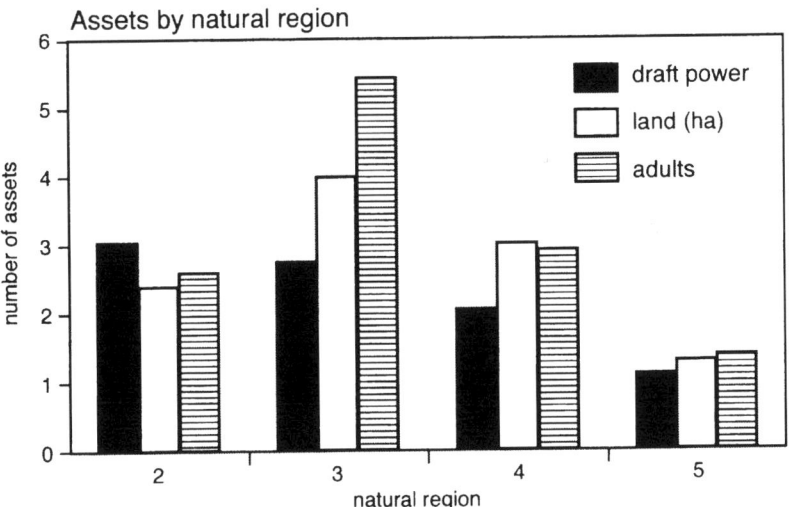

Draft power = number of animals; Land = hectares of arable land; Adults = number of adults living in the household. The number of draft animals in NR 4 is an average of the three sample areas in that zone, two of which are in the western, Ndebele-speaking region.

Figure 7.2   Distribution of assets per household by Natural Region

Region. Arable land per household is surprisingly low in NR 2, although this figure is consistent with that reported by Dalton et al. (1997) for nearby Mutoko. The largest measure of arable land per household was found in NR 3 (4.01 ha per household).[3] As expected, the smallest area was found in NR 5, with only 1.31 ha per household. Similar trends are seen for the number of adults per household, which may reflect the degree to which off-farm employment is sought as a substitute for income deficits associated with low farm productivity.[4] Assets were slightly lower for female-headed households, with the discrepancy largely due to differences in the number of draft animals owned, rather than area of arable land held.

2. Access to Climate Information

Access to seasonal climate information was widespread in 1997/98, as a result of broad media coverage of the strong El Niño event. In the survey conducted prior to the rainy season of 1997/98, over 90% of households in all three Natural Regions had heard a seasonal forecast from sources outside the village (Table 7.1). Neither level of household assets nor Natural Region significantly influenced access to information.

This was not the case in the following year, when a La Niña event occurred. In 1998/99, there was far less media attention given to the forecast and the numbers of households reported to have heard seasonal climate information prior to the start of the rainy season dropped to less than 50% when averaged across all survey sites. This decrease provided the opportunity to consider how location or wealth might influence access to information.

There was a significant correlation between hearing the forecast in 1998/99 and the survey site. However, there was no clear relationship between rainfall zone and forecast dissemination, since households across all three NR 4 survey areas reported having heard the forecast more often than those in NRs 2, 3 or 5 (Table 7.1). These results simply suggest that the distribution channels for seasonal climate forecasts are very uneven.

Access to climate information was positively correlated with level of assets in 1998/99 (Table 7.1). Fewer asset-poor farmers heard a climate forecast prior to the rainy season in 1998/99. There is also evidence of gender differences in terms of access to forecasts: 39% of male heads of household received information while only 26% of female-run households heard a forecast.

In 1997/98, the only significant source of forecast information was radio, and there was no relationship between source and Natural Region or level of household assets. In 1998/99, radio remained the most important source for all groups, but discussions with neighbors and community members were also important (Table 7.2). Agricultural extension services were generally not an important source of information, but they played a somewhat stronger role in the eastern Communal Areas (NRs 2, 3, and 4a) than elsewhere.

Table 7.1 Percentage households reporting having heard a seasonal forecast by a) Natural Region and b) level of assets

Natural Region	2	3	4a	4b	4c	4(avg)	5
1997/98	-	96	90	-	-	-	99
1998/99	12	36	55	37	52	48	20

Level of assets	1	2	3
1997/98	96	96	99
1998/99	26	35	45

Table 7.2 Sources of climate forecast information by Natural Region (percentage of households reporting having heard a forecast in 1998/99)

Source	Natural Region					
	2	3	4a	4b	4c	5
Radio	89	59	71	71	38	60
Discussions	0	30	20	29	56	40
Ag. ext.	11	4	15	0	3	0
Met. service	0	7	0	0	0	0
Newspaper	0	0	10	0	15	7
Television	0	4	5	7	5	13

Note: Percentages may add to greater than 100% because of multiple sources used.

When source of information is arranged by asset level (Table 7.3), it appears that discussions with neighbors are more commonly used as a source of climate forecasts by the wealthier set than by households in the two lower levels of assets.[5] Television also played a slightly greater role as an information source for wealthier households. The primary distinction between households headed by men and those headed by women was that discussions were a source for a larger percentage of women than for men (female: 47%, male: 29%). Also, radio was less commonly used as a source of climate information among female-headed households (47%, as compared to 65% for male-headed households).

Table 7.3   Source of climate forecast information by level of household assets (percentage households reporting having heard a forecast in 1998/99)

| | Asset level | | |
| Source | 1 | 2 | 2 |
	(n = 37)	(n = 53)	(n = 69)
Radio	57	51	71
Discussions	46	42	71
Ag. extension	0	8	7
Met. service	0	0	3
Newspaper	8	9	4
Television	0	6	9

Note: Percentages may add to greater than 100% because of multiple sources used.

3. Crop Management Strategies, 1997/98

In the 1997/98 season, farmers who had either heard the externally produced forecast or had a local forecast for a poor rainy season stated that they intended to use three main strategies to mitigate against low yields:

- change area planted;
- change crop or cultivars used;
- change planting date.

When the sample population was stratified by NR, the strategies relied upon most often and the number of strategies cited by any one farmer varied by zone (Table 7.4a). Changes in crop or cultivar were cited more than twice as frequently by farmers in NR 5 as in NRs 3 or 4. Changes in planting dates were also cited more often in the drier zones. Asset level, on the other hand, does not appear to have an influence on the strategies cited, as farmers in each category mentioned all three strategies roughly the same number of times (Table 7.4b). This is somewhat surprising, given that both area planted and planting date depend on amount of land and draft animals owned.

Table 7.4   Crop management strategies cited by farmers who expected a poor rainy season in 1997/8, based on either having heard the external forecast or believing local indicators. Presented by a) Natural Region and b) level of assets

Strategy	*a) Natural Region*			*b) Level of assets*		
	3	*4*	*5*	*Low*	*Medium*	*High*
Change area planted	52	58	62	61	57	55
Change crop or cultivar	39	24	88	58	48	62
Change planting date	50	61	80	61	58	64

The number of strategies mentioned is likely to reflect both flexibility and the degree of risk aversion. Only in NR 5, the driest zone, did the largest number of households in the group cite a combination of all three strategies (Table 7.5a). When stratified by level of assets (Table 7.5b), all groups cited single strategies most often. Only two percent or less of households cited a fourth strategy (change row spacing) to mitigate against poor rainfall.

Table 7.5   Percentage of households citing single or multiple crop management strategies in preparation for the coming rainy season in 1997/98. Presented by a) Natural Region and b) level of assets

Number of strategies cited	a) Natural Region			b) Level of assets		
	3	4	5	Low	Medium	High
1	54	63	21	41	49	40
2	39	31	38	36	30	40
3	4	7	41	24	20	18

4. Crop Management Strategies, 1998/99

More detailed data on intentions were collected in 1998/99, when the seasonal forecast projected an above-normal rainfall season associated with a La Niña year. The following analysis is restricted to the sub-population of households who either heard the external forecast, or expected a "good" rainy season based on traditional indicators. With regard to area intended for planting, in NRs 3 and 4 roughly the same percentage of households as in 1997/98 indicated that they would change area from their normal practice (Figure 7.3a). Far fewer farmers in NR 5 decided to change area planted (30 percent). However, among those households who did, all indicated that they would increase area to take advantage of the expected good rainy season. In zones three and four, the majority of farmers intended to increase area planted. In NR 2, where a "normal" rainy season is relatively wet, the majority indicated that they would decrease area planted, for fear of too much water.

When divided by level of assets rather than rainfall zone (Figure 7.3b) the percentage of farmers intending to change area planted is clearly related to household assets, but the majority of farmers in all three groups intended to increase area planted.

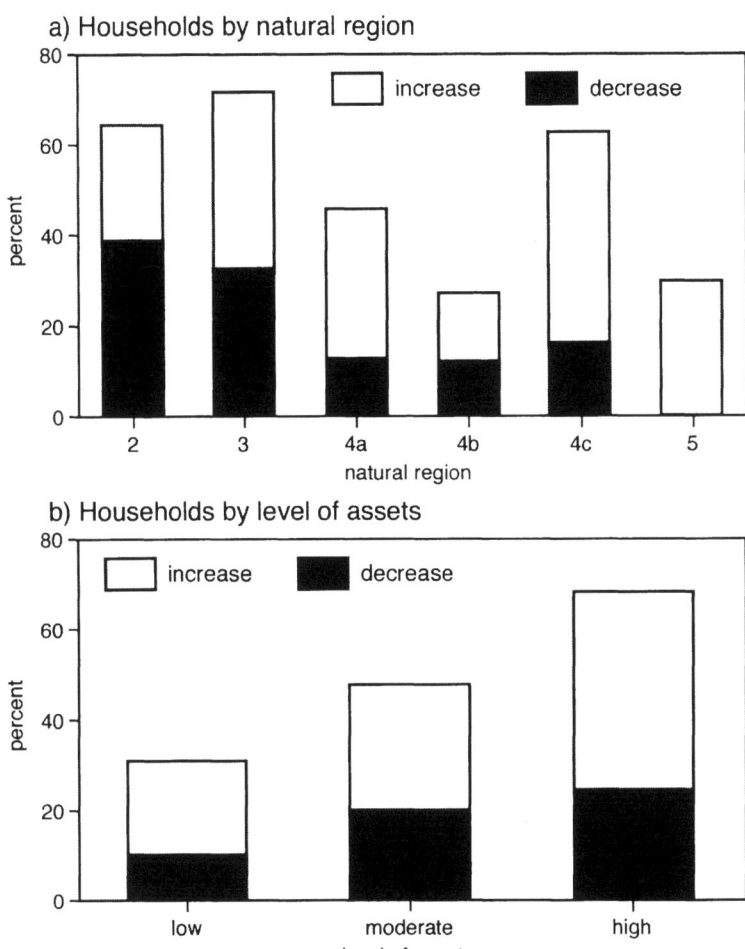

Figure 7.3   Percent of households planning to either increase or decrease area planted by more than 0.1 ha in 1998/99, compared to area normally planted. Presented by a) Natural Region and b) level of assets

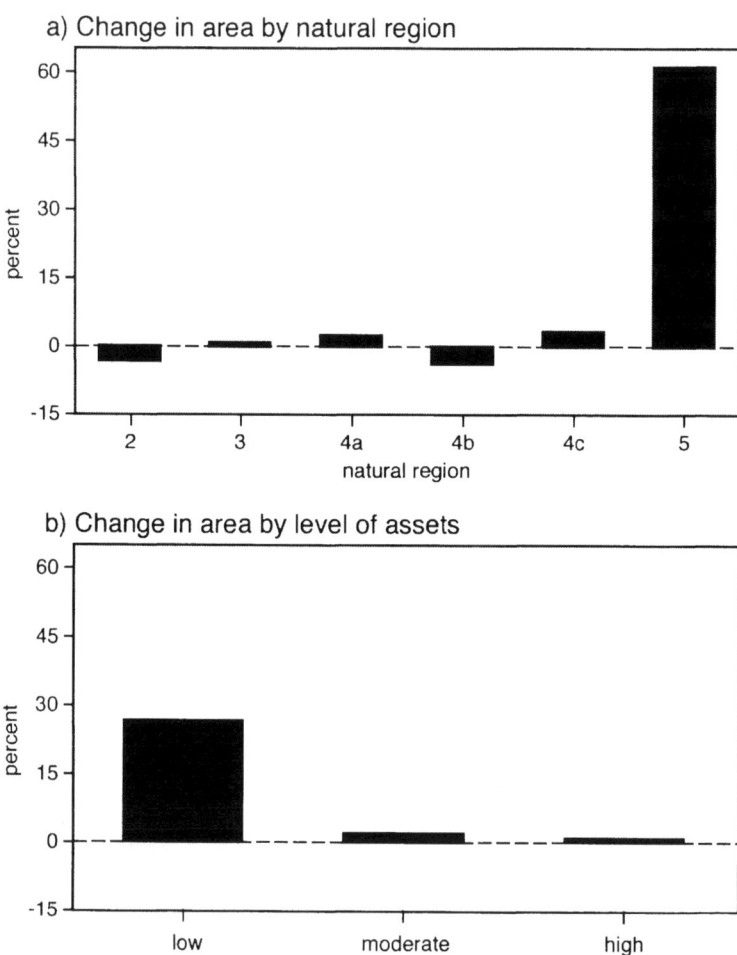

Figure 7.4   The percentage difference in total area planted comparing normal area planted to intentions for 1998/99. Presented by a) by Natural Region and b) level of assets

In Figures 7.4a and 7.4b, the percent change in area intended to be planted is shown by NR and level of assets. Because all of the farmers intending to change area in NR 5 planned to increase area, the average change in area is quite high (an increase of 61%) (Figure 7.4a). Similarly, although fewer farmers with low levels of assets intended to change area planted compared to the wealthier farmers, the majority of them intended to increase area, so the average change is positive (Figure 7.4b).

The percentage of farmers intending to change crop type or cultivar was assessed with respect to maize (which requires a good rainfall season to yield well) and two representative small grains: pearl millet and sorghum. When analyzed by Natural Region, there was no change in the number of farmers planting maize in NRs 2 or 3 – all farmers in the sub-population of those expecting a good rainy season normally plant maize and intended to do the same that year (Table 7.6a). However, in the driest zone, NR 5, there was a 25% increase in the number of farmers planting maize. Maize is risky in NR5, but can yield better than the small grains in a moderately wet year. Conversely, 58% of farmers in NR 3 who normally plant small grains (pearl millet or sorghum) decided not to plant them that year. These crops are less preferred to maize, but are usually planted as insurance against the likelihood of a poor rainy season because they are drought tolerant. Some farmers in NRs 4 and 5 also decided to leave millet and sorghum out of their crop mix in 1998/99 (Table 7.6a).

Table 7.6 Percent change in the number of farmers planting maize or the small grains pearl millet or sorghum, sorted by a) Natural Region and b) level of assets

	a) Natural Region				*b) Level of assets*		
	2	*3*	*4*	*5*	*1*	*2*	*3*
Maize	0	0	3	25	9	-2	4
Sorghum or pearl millet	0	-58	-2	-12	-6	-9	-6

Although seed for most of the small grains are saved at the village level, and thus are inexpensive and readily available, maize seed is purchased and represents a considerable proportion of the costs of planting. One might therefore expect wealth to influence decisions to plant maize. However, when the population is sorted by level of assets, very little distinction can be seen between asset levels with respect to decisions on planting maize or the small grains (Table 7.6b).

Nearly all farmers expected to change planting date in response to the forecast, and in all groups the expected date was slightly later than normal. Normally, most farmers prefer to have planting completed just after the rainy season begins between mid October and mid November. A representative distribution of dates, showing expected planting dates for the high asset level group of households, is shown in Figure 7.5. In addition to showing a slight shift to later planting, Figure 7.5 also shows a narrowing of the window of planting, which was the case for all groups (data not shown). This may have partially been a result of having an advanced indication as to when they would be able to complete planting.

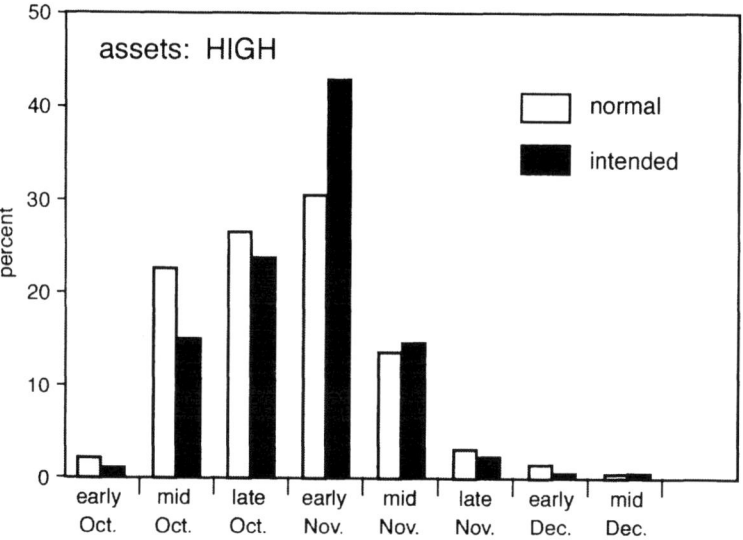

Figure 7.5 Distribution of farmers' planting dates, normal and expected, in 1998/99 for the sub-group of farmers in the top third of assets

Discussion and Conclusions

The research presented here suggests the following: 1) assets do not appear to hinder use of a forecast as much as was expected, but 2) access to information *is* likely to be influenced by wealth, and 3) forecasts of a good year may have as much or more value than poor years.

The primary influence of asset level was found to be on access to climate information. Clearly, when an information campaign is moderate as was the case in 1998/99, the greater the level of assets, the more likely the household is to receive news of the forecast from outside the community (e.g., via radio and television). Wealthier farmers were more likely to also have regular contact with agricultural extension officers and even local meteorological experts in their communities. This was not the case in 1997/98, however, due to the strong coverage of the "El Niño of the century." While traveling in the rural areas that year, one heard the word "El Niño" sprinkled though out conversation in Shona and Ndebele. This suggests that with a large enough information campaign, everyone potentially has access to information.

In addition to determining who receives the message, access to assets also assists in farm decision-making strategies. Asset level was found to be important for decisions regarding changing area planted based on forecast information. A greater number of farmers in the higher assets groups planned to change area planted in 1998/99, compared to the area they normally plant. No distinction between asset levels was found in 1997/98 regarding the decision to change area planted. This is somewhat surprising, because in expecting a poor year, the likely change would have been to reduce area, which should be more of an option for a household that already produces an excess than for one producing close to its annual consumption needs.

Level of household assets did not seem to play a role in determining either crop choice or planting date decisions in the context of the forecasts. Although farmers with greater draft animal ownership, which is a key component of the asset variable used here, generally plant earlier than farmers without primary access to draft power, the changes in planting date reported here did not vary as a function of asset level. All farmers expected to plant slightly later than normal in 1998/99. This may have been a response to the expectation of a good year, which is usually perceived to be associated with a long rainy season and the lack of the mid-season dry spell (Phillips, Cane, and Rosenzweig, 1998).

Instead of assets, the biophysical environment (e.g., agroecological zone) plays a strong role in determining crop-management decisions. In 1998/99, there was a general shift away from the small grains and into maize given the expectation of a wet year. That shift was tempered by zone in that large numbers of farmers in the wettest areas decided not to plant pearl millet or sorghum, while only slightly fewer in the driest zone left the small grains out completely. Given the risks normally associated with planting maize in drier areas, the expectation of an above normal rainfall season led to a more than doubling of the number of farmers there who intended to plant maize. This management strategy is one likely to lead to very real benefits, if forecast skill is appropriately taken into consideration.

Particularly in the more marginal rainfall zones, crop choice is critical in determining production, and yields of the major grains vary considerably depending on rainfall. Indeed, it has been shown that even by making relatively small changes in the proportions of area devoted to maize or the small grains, based simply on ENSO phase, farm-level production in the Communal Areas in Zimbabwe can be significantly increased (Deane, 1997). Other work has shown a large degree of flexibility in proportions and crops planted among Communal farmers in Zimbabwe (Scoones, 1996a, 1996b), supporting the idea that shifts in crop choice are a viable strategic response to seasonal forecasts.

Finally, these results provide evidence of the benefits of using a forecast for an above-normal rainy season to improve production among a normally risk-averse population. Although forecast uncertainty needs to be communicated appropriately, so as not to increase risks for this group, the results support the idea that forecasts for good years may lead to benefits commensurate with forecasts for poor years for at-risk small farmers. However, because area planted is a management strategy constrained by the amount of available land, benefits will more likely accrue to those who have large areas available. It is important to monitor potential inequities that may develop as a result of forecast availability.

In conclusion, the level of household assets, on the one hand, was found to be less important than previously assumed, at least among the relatively narrow range of wealth found in this sample of Communal Area farmers. Agroecological zone, on the other hand, appears to be an important factor in determining how climate information is used in crop management.

Although household assets are related to likelihood of hearing a forecast, this inequity can be addressed through improved information campaigns, with special attention paid to giving forecasts of good years equal weight as those of years in which below normal rainfall is expected.

Overall, these results show that even poor farmers with limited resources can respond to forecast information, potentially contributing positively to food security in Africa. Farmers in marginal rainfall zones potentially have the most to gain from quality information, but will need to be well informed regarding uncertainties inherent in seasonal forecasts.

Notes

1 Natural Region 1, on the east escarpment along the Mozambique border, is spatially limited and dominated by export crops such as coffee and tea.

2 Decreases in herd size as a result of the droughts of the early 1990s have contributed to a decrease in manure applications, as cattle populations have been slow to recover (Carter and Murwira, 1995; Scoones, 1996b).

3 This relatively large area is likely to be a function of our particular sample, as figures for area cropped per household from the NR 3 zone of Save CA were found to be 2.9 ha in the early 1980s (Campbell, Du Toit, and Attwell, 1989) and 2.5 ha in the early 1990s (Dalton, Masters, and Foster, 1997).

4 Shumba (1994) notes that more than half of the household income in a sample of households from Chivi CA in NR4 is derived from off-farm activities, which represents a much greater proportion than that found in NR2.

5 Although differences were not significant between asset levels, contact with both agricultural extension services and local meteorological services was higher for households with greater assets.

References

Campbell, B.M., Du Toit, R.F., and Attwell, C.A.M. (1989), 'The Save Study. Relationships Between the Environment and Basic Needs Satisfaction in the Save Catchment, Zimbabwe', University of Zimbabwe, Harare.

Carter, S.E., and Murwira, H.K. (1995), 'Spatial Variability in Soil Fertility Management and Crop Response in Mutoko Communal Area, Zimbabwe', *Ambio*, vol. 24, pp. 77-84.

Dalton, T.J., Masters, W.A., and Foster, K.A. (1997), 'Production Costs and Input Substitution in Zimbabwe's Smallholder Agriculture', *Agricultural Economics*, vol. 17, pp. 201-209.

Deane, D. N. (1997), 'Climate, Crops and Cash: An Investigation into the Impact of El Niño-Southern Oscillation (ENSO) Forecasts on Communal Farmers in Zimbabwe', PhD. Thesis presented at Williams College, Williamstown, Mass.

Huchu, P., and Sithole, P.N. (1994), 'Rates of adoption of new technology and climatic risk in the communal areas of Zimbabwe', E.T. Craswell and J. Simpson (eds), *Soil Fertility and Climatic Constraints in Dryland Agriculture*, Proceedings of the ACIAR/SACCAR workshop held at Harare, Zimbabwe, 30 August - 1 September 1993. ACIAR Proceedings No. 54, pp. 44-50.

Kinsey, B., Burger, K, and Gunning, J.W. (1998), 'Coping with Drought in Zimbabwe: Survey Evidence on Responses for Rural Households to Risk', *World Development*, vol. 26, pp. 89-110.

Masters, W.A. (1994), *Government and Agriculture in Zimbabwe*, Praeger, London.

Ministry of Lands, Agriculture and Rural Resettlement, (1990), 'The Second Annual Report of Farm Management Data for Communal Area Farm Units, 1989/90 Farming Season', Farm Management Research Division, Harare.

Pfaff, A., Broad, K., and Glantz, M.G. (1999), 'Who Benefits from Climate Forecasts?' *Nature*, vol. 397, pp. 645-646.

Phillips, J.G., Cane, M.A., and Rosenzweig, C. (1998), 'ENSO, Seasonal Rainfall Patterns, and Simulated Maize Yield Variability in Zimbabwe', *Agriculture and Forest Meteorology*, vol. 90, pp. 39-50.

Phillips, J.G., Makaudze, E. and Unganai, L. (2001), 'Current and Potential Use of Climate Forecasts for Resource-poor Farmers in Zimbabwe', in American Society of Agronomy Special Publication #63. *Impacts of El Niño and Climate Variability on Agriculture*, pp. 87-100.

Scoones, I. (1995), 'Investigating Difference: Application of Wealth Ranking and Household Survey Approaches among Farming Households in Southern Zimbabwe', *Development and Change*, vol. 26, pp. 67-88.

Scoones, I. (1996a), 'Crop Production in a Variable Environment: a Case Study from Southern Zimbabwe', *Experimental Agriculture*, vol. 32, pp. 291-303.

Scoones, I. (1996b), Hazards and Opportunities. Farming livelihoods in Dryland Africa: Lessons from Zimbabwe. Zed Books, London.

Shumba, E.M. (1994), 'Constraints in the Smallholder Farming Systems of Zimbabwe', in E.T. Craswell and J. Simpson (eds), *Soil Fertility and Climatic Constraints in Dryland Agriculture*, Proceedings of the ACIAR/SACCAR workshop held at Harare, Zimbabwe, 30 August - 1 September 1993. ACIAR Proceedings No. 54, pp. 69-73.

8 Climate Forecasts in Mozambique: An Economic Perspective

CHANNING ARNDT, MELANIE BACOU, AND ANTONIO CRUZ

Introduction

Climate variations can have major implications for humankind. In developed economies, the scale of economic activity in climate-sensitive sectors and related upstream and downstream activities implies that even relatively small climate anomalies can generate significant losses in absolute dollar terms. In developing countries, the scale of economic activity is much smaller, and the absolute dollar value of costs of climate disruptions is commensurately lower. In relative terms, however, the costs of climate variability in developing countries can be enormous, due to a relatively high importance of climate-sensitive industries in the overall economy. In addition, the human cost of significant variations in climatic conditions can be very high due to poverty and the related lack of capacity to cope with climate shocks once they have occurred.

Reliable climate forecasts provide one potential means for reducing the economic effects of climate variability. Recent improvements in medium-range forecasts of several months to one year have enhanced this potential considerably. Nevertheless, while a growing body of research suggests that reliable climate forecasts should be valuable, the magnitude of potential economic gains, as well as the sources of those gains, remains poorly understood.

In this chapter, the economics of climate forecasts are considered, with particular attention to the case Mozambique. To provide a context for understanding the economic consequences of climate variability and the potential benefits of seasonal forecasts, a wide range of literature is reviewed. Much of this literature is based on analyses of developed economies, particularly the United States. The climatic and economic situation in Mozambique is then described, and the results from a modeling study that examined the economic value of climate forecasts in Mozambique are presented. These results point to the agricultural marketing system as a major potential source of gains from the application of climate information. Following up on this result, a rapid appraisal survey of the marketing system in Mozambique was carried out to identify potential options for reducing domestic marketing margins.

Economic Perspectives on Climate Variability and Forecasts

Agricultural production is closely dependent upon climatic factors, including climate variability. El Niño Southern Oscillation (ENSO) variables explain about 15 to 35% of global average yield variation in coarse grains, wheat, and oilseeds over the past 40 years (Ferris, 1999). Consequently, there has been a substantial amount of research on the economics of climate variability and climate forecasts. For example, the economic consequences of El Niño (warm phase) and La Niña (cold phase) events for primary agriculture in the United States have been estimated (Adams et al., 1998). Both phases represent departures from normal conditions, and both phases inflict economic costs. Costs of USD 1.5 to 1.7 billion (1990 dollars) for the El Niño phase and USD 2.2 to 6.5 billion dollars for the La Niña phase have been estimated, with the variation in figures related to the severity of the event (Adams et al., 1998). While these numbers seem large in absolute terms, they are quite small relative to the size of the U.S. economy. In fact, even the largest figure (adjusted for inflation) amounts to less than one tenth of one percent of U.S. GDP in 1999. In other words, if one uses agricultural output as an imperfect proxy for climate variability, the implications of climate variability for food systems in the United States appear to pale relative to the size of the overall economy.

Implications of climate variability for food systems, however, extend well beyond the farm, such that estimates focusing purely on agricultural production can significantly understate the true costs of climate variability. Reduced agricultural output, for example, can have repercussions on food processing sectors, which significantly magnifies the economic impacts of agricultural production declines. Even so, an analysis of data from 1975 to 1996 (Council of Economic Advisors, 2001), suggests that variations in agricultural output in the United States (ostensibly due in large part to climate variability) have not had a statistically recognizable effect on overall output growth rates. This lack of effect stems from the relatively small share of primary agriculture in terms of GDP (about 1.5% in the 1990s), the great diversity of U.S. agriculture which leads to a high degree of stability in aggregate production, and the quality of infrastructure and transport services, which cushions the impact of local production shortfalls on downstream industries such as food processing.

In short, losses due to climate variability appear large in the United States in terms of absolute value, but small relative to the size of the economy. The same conclusion can probably be applied to many other developed countries. The opposite situation, however, holds true for developing countries such as Mozambique. In such economies, the absolute values may be minor because the overall size of the economy is small, but the relative effect may nevertheless be large. Indeed, the total value of agricultural output in Mozambique amounts to less than half of the lowest cost of variability figure estimated by Adams et al. (1998)–that is, less than half of USD 1.5 billion. Nevertheless, the agricultural sector directly or indirectly provides employment or livelihoods for about 80% of the economically active population and contributes roughly 30% of GDP (National Institute of Statistics, 2000).

Benson and Clay (1998) examined the economic impacts of drought for countries in sub-Saharan Africa and found that drought has a significant impact on economic growth. This impact can be attributed to two factors: First, it is related to the high share of primary agriculture in GDP. The analysis shows that the larger the primary agricultural sector, the larger the first-order effect of drought. Second, it can be attributed to the relatively high economic importance of downstream linked industries, such as food processing and textiles. The strength of this effect depends on infrastructure and transport links, which are characteristically low in most African countries. Droughts and other disruptions in primary production can severely affect downstream industries due to a lack of primary product available in local markets. Furthermore, poor transport links can prevent regional or international supplies from filling the void at a reasonable cost.

What is the potential role of climate forecasts in mitigating these costs? In order to be valuable, a climate forecast must induce economic agents to take actions to improve economic conditions, above and beyond what would have been done in the absence of a forecast. Most analyses of the economic benefits of climate forecasts have taken place in the context of developed countries, and much attention has focused on agriculture and related industries. An analysis at the farm level was undertaken (Mjelde et al., 1993), and positive values of forecasts for Illinois corn farmers were recorded. The same study also pointed out that a less reliable but earlier forecast may be more valuable than an accurate but late forecast. Solow et al. (1998) extended the farm-level analysis to cover farmer reactions for all of the United States. They found the expected value of a perfect ENSO prediction (combined with perfect farmer reactions) to amount to USD 323 million per year. A climate forecast of more modest skill reduces the expected benefit by about 25%.

At the marketing system level, Ker and McGowan (1999) examined the implications of climate forecasts for U.S. crop insurance, emphasizing interactions between forecast information, insurers, and current crop insurance subsidization policies of the U.S. government. Focusing on the particular case of Texas wheat farmers, they found that if insurers simply factor in observations on sea-surface temperatures in the equatorial Pacific, the added information could significantly influence the expected value of an insurance policy. In short, climate forecast information permits insurers to secure the most desirable insurance policies and leave the government with the least desirable policies, with the result that insurance companies earn large profits while the government suffers large losses. While this particular set of policy/forecast interactions is unlikely to enhance economic efficiency, it does provide the prospect, under an appropriate policy environment, of more informed decision-making on the part of insurers and farmers with associated, and potentially significant, improvements in efficiency.

The value of ENSO forecasts for U.S. corn storage has also been examined (McNew, 1997). Inter-annual commodity storage occurs primarily to smooth consumption levels in the face of variable production. Specifically, storage can help to maintain consumption despite a production shortfall. Forecast information can facilitate this process by indicating the likelihood of a production shortfall. If a production shortfall is highly unlikely, there is little need to hold inter-annual storage. If a production shortfall is likely, storage volumes can be increased in order to smooth consumption over a longer period of time. Overall, forecast information has the potential to reduce the average volume of inter-annual storage, while achieving the same or improved levels of consumption smoothing. Based on this idea, McNew (1997) estimated that a perfect ENSO forecast could reduce the average volume of storage, with benefits adding up to nearly $240 million annually.

Gains from corn storage alone could thus contribute to as much as 75% of the gains from farmer responses for the entire agricultural sector (assuming a perfect forecast in both cases). It is worth highlighting that, assuming the relative magnitudes are reasonably correct, the gains from forecast information in storage of all crops (e.g., adding in the gains from increased efficiency in storage of soybeans, wheat, etc.) should easily surpass the gains from farmer responses on the production side. Gains from responses in input supply (crop insurance being only one possibility), transportation, and processing would further highlight the importance of marketing sector agents in realizing value from climate forecasts.

Forecasts and their Economic Value: the Case of Africa

An integrated methodology for examining the economics of climate forecasts, with a particular focus on the agricultural sector in Africa was introduced by Arndt, Hazell, and Robinson (2000). Simulation modeling, as a means of generating *with forecast* and *without forecast* scenarios was proposed, and three levels of analysis were identified. The first involved farmer reactions as described by Adams et al. (1998). The second involved the reactions of agents in the marketing system. The marketing system encompasses input supply and output purchase, storage, transport, and transformation/processing. A third level of analysis provided an economy-wide perspective, which permits one to consider how these shocks and reactions interact and add-up. This last level is important in the African context due to the large share of agriculture and marketing activities in overall GDP.

Input supply is one area where agents in the marketing sector in Africa have significant scope for responding to and benefiting from forecasts. For example, in the Sahel region of Africa, aggregate demand for agricultural inputs, such as fertilizer and herbicides, varies dramatically with the aggregate level of rainfall (Arndt, Hazell and Robinson, 2000). In years with above-average rainfall, total demand for agricultural inputs (at a constant price) is well above total demand in years with below-average rainfall. Unfortunately, Sahelian countries are both highly dependent on imports of either the inputs themselves or their component chemicals. It is therefore very difficult to moderate aggregate agricultural input availability with realized climate outcomes. Total input supply is effectively determined prior to realized demand. Climate information could play a key role in more efficiently matching aggregate input supply with aggregate input demand, at least on the average.

In attempting to determine sectors with high potential for realizing value from climate forecasts, the key issue is the scope for actions, which can create significant differences between economic outcomes in the with forecast and without forecast scenarios. While farmers can modulate input use, crop and crop-variety choice, and other production practices in accordance with a climate forecast, the unfavorable climate conditions cannot be avoided, such that much of the cost of drought is incurred in both the with forecast and without forecast scenarios.

A focus on the marketing system is also consistent with current climate prediction capabilities with regard to the scale of analysis (Arndt, Hazell, and Robinson, 2000). Harrison and Graham (1998) point out that the skill of atmospheric models is maximized at regional scales, typically on the order of 105 [square] kilometers or more. While some marketing sector agents operate at very local levels, many others, including major input suppliers, operate at the regional level, where climate prediction skill is highest. Dissemination of forecast information to the marketing sector is also simpler, particularly in a developing country context, since marketing agents are relatively limited in number. Finally, in developing countries, marketing sector participants are likely to have higher education levels on average than farmers. Hence, in these early stages of experience with climate forecasts, the task of communicating the forecasts to marketing agents is likely to be substantially less daunting than the task of communicating the forecast to subsistence farmers.

Economic Impacts of Forecasts in Mozambique

Climatic Context

Mozambique is a tropical country located in southeastern Africa. It is bordered by Tanzania to the North; Malawi, Zambia, Zimbabwe, and Swaziland to the West; and South Africa to the South. It has a 2,515 km (1,572 miles) coastline (NIS, 1998), which is significantly longer than the coastline of the western United States. There are two main climate seasons: a warm and humid season extending from November to April, and a cooler and drier season extending from May to October.

Average temperature and rainfall in Mozambique varies according to location. Three main climate zones can be identified (MINED, 1986). The first climate zone includes the provinces of Cabo Delgado, Niassa (excepting mountain zones in the southwest), Nampula, and Zambezia plus a coastal strip of Sofala province extending about 50 km inland (see Figure 8.1). This zone is tropical and humid, with monthly maximum rainfall averaging 240 to 300 mm and average temperatures ranging from 20 to 30°C.

The second climate zone includes the Zambezi River valley, north of Manica province and lower half of Tete province, and the interior region from the Save River valley to the southern end of Mozambique. The Save River separates Sofala and Manica provinces on the center and Inhambane and Gaza provinces in the south. The climate is tropical and dry, with temperatures varying from the 20s to low 30s in degrees Celsius. Monthly maximum rainfall averages 160 mm. Annually, it rains less and the average temperature is higher than in the first zone.

The third climate zone includes the central region in Manica province, from Sofala to the border with Zimbabwe, the northern part of Tete province, the Gurue region in Zambezia and the mid-southwestern portion of Niassa province. It rains considerably during the warm season, but becomes particularly dry in the cooler season. The temperature averages 20°C, which is lower than in the other two climate zones and can be attributed to higher altitudes.

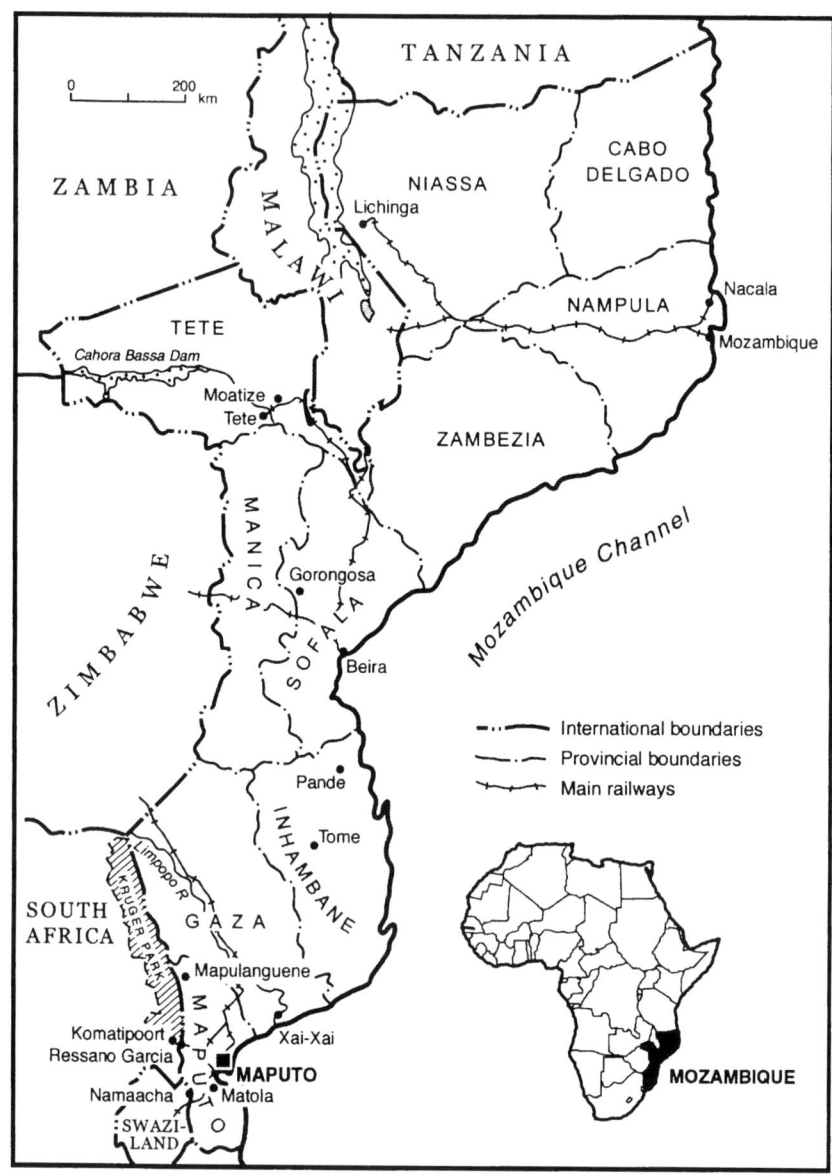

Figure 8.1   Map of Mozambique

Agricultural production is highly dependent upon the quality and duration of the rainy season. Annual precipitation levels, however, are subject to large inter-annual fluctuations. Regional variations are important as well. Figures 8.2 and 8.3 show Mozambique's average temperatures and precipitation by province. Average rainfall levels are derived from 30-year observations (1931 to 1960). Mountains and high plateaus in the western and northern provinces can receive up to 2000 mm yearly, while erratic rainfall is to be expected in the southern provinces.

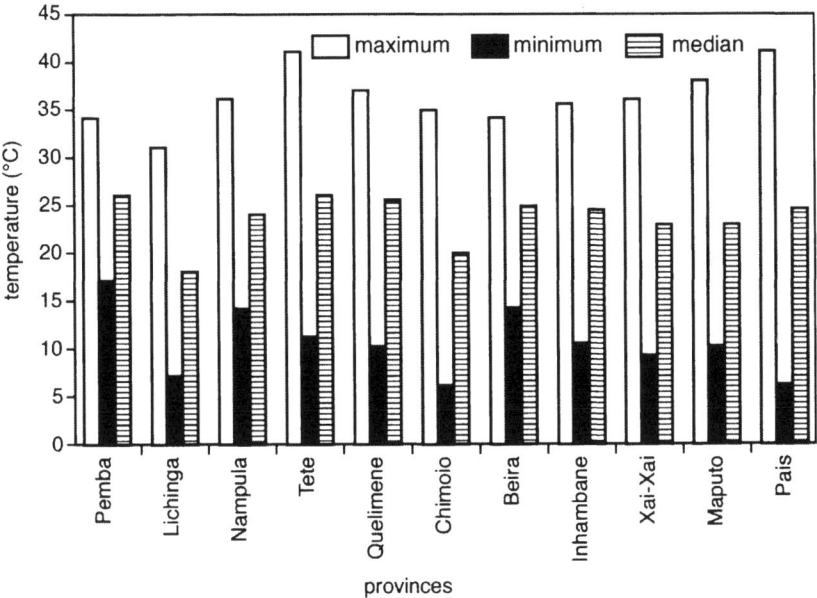

Source: Instituto Nacional de Meteorologia, 1996

Figure 8.2   Long-term average temperatures

If rainfall varies from the norm, both in terms of total precipitation and in timing, food security is affected. Data from Mozambique illustrate this nexus. Between 1961 and 1993, average annual rainfall for the whole of Mozambique was 1032 mm. As shown in Figure 8.4, grain yields fluctuated widely during that time. Within the grain category, maize exhibited the highest yield variability. Over the whole period, maize yields fluctuated from 0.2 to 1.1 tons per hectare. These wide fluctuations parallel the changes in annual rainfall.

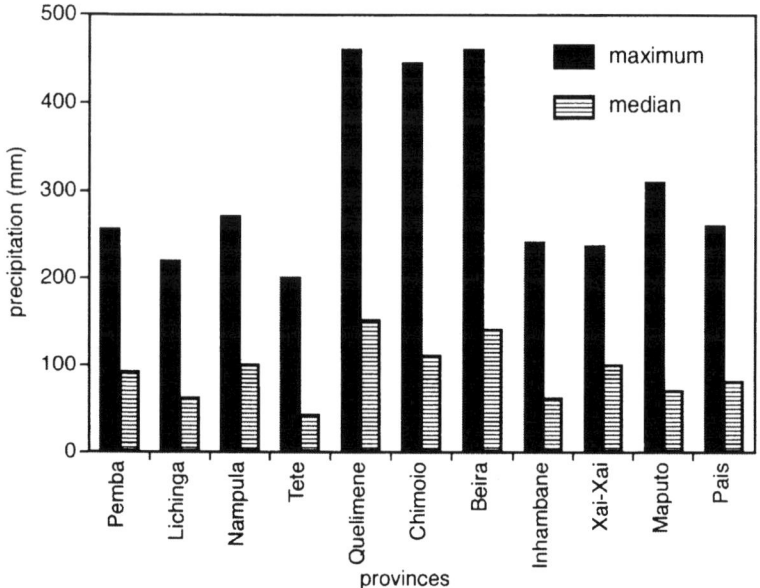

Source: *Instituto Nacional de Meteorologia, 1996*

Figure 8.3   Long-term average precipitation

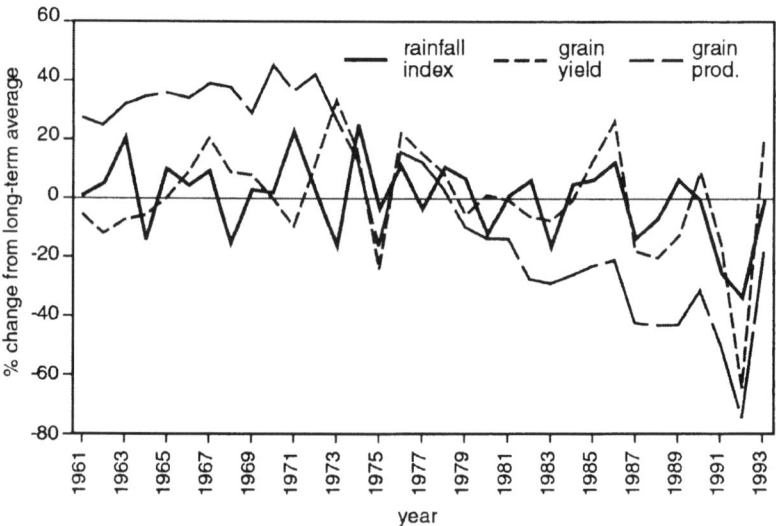

Figure 8.4   Influence of inter-annual rainfall variability on grain yields and production, 1961–1993

Mozambique has experienced many severe droughts since independence in 1975. During the cropping years 1982/83, 1986/87 and 1991/92, maize yields fell on average by 40% (MERISSA, 1998), and by up to 85% in the center of the country (Tete and Manica provinces). Such dramatic events caused large-scale food deficits, hunger, famine, flight of people and animals, and diseases. National and international trade flows were also affected. This had far-reaching economic consequences for the country, such as reductions in foreign exchange earnings, increased food imports, worsening national debt burden, which in turn threatened the country's capacity to cope with the negative impacts of drought.

Structural drought risks are a major concern for agriculture in the southern provinces of Inhambane, Gaza, and Maputo, with the exception of the coastal districts. Drought also affects the central province of Tete. Floods are very common along the Zambeze valley. Some districts are faced with both drought and flood risks (Matutuine, Chibuti, Moatize, Cahora Bassa). Inter-regional trade offers potential for mitigating the negative effects of extreme climatic conditions in the country. Poor north-south transportation infrastructure, however, has impaired such mitigation strategy in the past.

In Mozambique, ENSO teleconnections appear to be particularly strong. Figure 8.5 shows the variations in Mozambique's national rainfall index in relation to a Multivariate ENSO Index since 1961. The figure indicates that floods and droughts do not occur at random. Instead, inter-annual rainfall patterns closely follow the cycles of warm and cold ENSO phases. Under these conditions, early and reliable ENSO forecasts offer potential for the economy to better prepare and mitigate the negative effects of climatic variability.

Economic Context

Mozambique gained independence from Portugal in June 1975. Immediately after independence, a centrally planned economic system was adopted (Arndt, Jensen, and Tarp, 2000). Frelimo, the party in power, chose this economic system as a strategy to develop the economy consistent with its ideals of a fair society. Support from socialist countries was expected to supplant the economic role of the quarter million Portuguese and thousands of Mozambicans who left the country in the aftermath of independence. At the time of independence, extremely few Mozambicans had a college education, and overall rates of educational attainment were low, even by African standards.

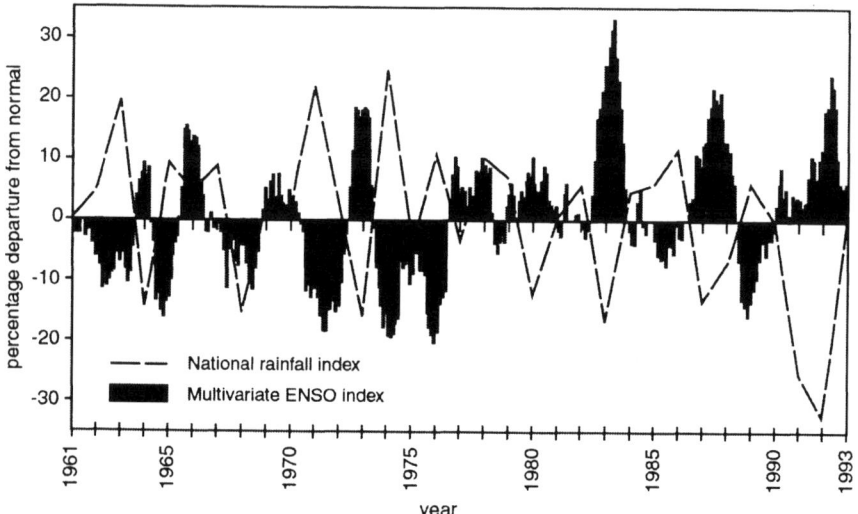

Sources: Gommes and Petrassi, 1994 and NOAA Climate Diagnostic Center, 1999, respectively.

Figure 8.5   National Rainfall Index (line) and Multivariate ENSO Index (shaded), 1961–1993

Given the lack of indigenous human capital, the Portuguese exodus immediately following independence posed an enormous economic challenge. Those who left for Portugal had been running most of the businesses and government services in Mozambique (Tibana, 1994). Faced with severe constraints on the availability of human capital, the economy experienced a crisis in the immediate post-independence period, but recovered slowly until 1981. From 1982 onwards, the economy entered a long decade of depression (for example, see grain yields and production in Figure 8.4). Failed economic policies, internal armed opposition (Renamo), the international oil crisis, confrontations with former Rhodesia and South Africa (both of whom provided support to Renamo), droughts in 1982/83 and 1991/92, and the collapse of the socialist system based in the USSR all contributed to this economic crisis (Arndt, Jensen, and Tarp, 2000).

In 1983, the Mozambican government sought economic support from western countries, and in 1987 it implemented a structural adjustment program with the support of the International Monetary Fund, the World Bank, and western donor countries. Although the government implemented some standard macroeconomic policy reforms, and donors financed part of the country's imports, economic conditions failed to improve due to the continuing internal armed conflict. Instability also effectively prevented implementation of a microeconomic reform program. Following the drought of 1991/92, Mozambique earned the unwanted label of "poorest country in the world."

Frelimo and Renamo signed a peace accord in 1992. The accord was implemented, and the first multiparty elections in the country's history took place in October 1994. With internal social and political stability, economic conditions began to improve, albeit from a dismal base. Peace permitted further macroeconomic reforms and a fairly vigorous microeconomic reform program. These reforms, combined with favorable weather conditions, significant external assistance, a good international economic environment, and a natural tendency for rebound after a prolonged period of depression, have led to reasonably good economic performance. In the second half of 1990s, the economy grew at rates higher than 7% annually on average (NIS, 2000).

More than 70% of the Mozambican population lives in rural areas (NIS, 1999). Rural inhabitants are almost completely dependent on agriculture for income. Typically, rural inhabitants are smallholders operating subsistence farms. The main function of this family-based agriculture is to provide its members with food and other elementary subsistence goods. Isolation from markets is a feature of many of these households. Home consumption (valued at producer prices) represents about 44% of the value of total consumption for rural households (Arndt, Hazell, and Robinson, 2000). Some smallholder producers are integrated into the market through the sale of surplus food crops and cash crops and the provision of seasonal labor. Finally, migration of some family members off farm, particularly to South Africa, provides income in the form of remittances, particularly in southern regions.

According to national accounts' data from the National Institute of Statistics (NIS), the primary, secondary, and tertiary sectors[1] accounted respectively for 32%, 16%, and 52% of GDP in 1997 (NIS, 1997). The relatively small size of the secondary sector has been a policy concern, and the government has tried various programs to revitalize this sector, in particular the manufacturing sector. Primary production, with agriculture represents the largest share, and along with services continues to dominate the economy.

In services, domestic trade, primarily of agricultural products, is particularly important accounting for about a quarter of GDP. A high level of transaction costs due to high risks in commercial transactions, a limited network of commercial agents, poor transport infrastructures, large distances, and poor communications networks, form the primary explanation for the large share of resources devoted to domestic commerce. Reductions in these transactions costs (the flip side of increased efficiency in marketing) have been shown to generate substantial economy-wide gains (Arndt, Hazell, and Robinson, 2000).

Two factors explain the relatively large size of domestic trade. First, Mozambique is highly dependent on foreign aid to finance its imports. Basically, the economy receives a large inflow of commodities, which are shipped to various points in the country through the commercial chain and sold to customers. Second, there is a high level of transaction costs due to high risks in commercial transactions, a limited network of commercial agents, poor transport infrastructures, large distances, and poor communications networks. Most of these costs are concentrated in agriculture and downstream industries. Reductions in these transactions costs (the flip side of increased efficiency in marketing) generated substantial economy-wide gains (Arndt, Hazell, and Robinson, 2000).

The next section examines the benefits of climate forecasts derived from assumed responses by farmers and marketing sector agents. The relative size and distribution of the benefits from these responses are compared. The section thereafter then takes a preliminary look at the composition of marketing margins and the scope for marketing sector agents to respond to forecast information.

Climate Forecasts in Mozambique: A CGE Approach

A first step analysis of the impacts of drought and the potential role of drought forecasts has been undertaken using a computable general equilibrium (CGE) approach for Mozambique (Arndt and Bacou, 2000). This represents an economy-wide approach, the most aggregate level of analysis as described by Arndt, Hazell, and Robinson (2000). Using an economy-wide approach, one can examine the linkages and feedback effects between farmer responses, responses of the marketing sector, and responses of the non-agricultural sector. As emphasized above, poor or favorable climate can have a major macroeconomic impact when agriculture and agricultural processing represent a substantial portion of economic activity. The economy-wide approach permits one to account for these spillovers and interactions.

The attraction of the CGE approach lies in its completeness (Arndt, Hazell, Robinson, 2000). Many of the relationships defined in a CGE model are accounting relationships, which must hold both in the model and in an economy. For example, CGE models include the basic accounting constraints listed below:

- The value of imports cannot exceed total foreign exchange availability (including foreign borrowings);
- Total consumption of a good cannot exceed total supply;
- Investment cannot exceed total savings;
- Government spending must be financed;
- Households must respect their budget constraint;
- Production requires real resources;
- Factors of production (e.g., labor and capital) are in finite supply;
- Demand for factors of production by productive activities cannot exceed factor supply; and
- Households (and potentially government) own and receive revenue streams from factors of production.

Although none of these propositions are in dispute, they serve to circumscribe, sometimes surprisingly tightly, the range of feasible economic outcomes. In addition to this accounting framework, the economic behavior of agents in the system is specified. In classic CGE models, consumers maximize utility of consumption and producers maximize profit in a perfectly competitive environment. Government behavior is exogenous and time is not explicitly treated (e.g., the model is static). Current applications often depart significantly from this classical standard. Recent CGE applications explicitly incorporate risk-reducing behavior, specify labor-leisure tradeoffs, examine the implications of imperfect competition, and/or incorporate dynamics (Francois and Reinert, 1997).

The CGE approach is particularly valuable for assessing the impacts, including distributional effects, of large shocks. When shocks are large, interaction and feedback effects can be very important. For example, consider a developing economy that exports a cash crop and imports food crops and manufactured goods. With agriculture accounting for a high share of GDP, a drought can be expected to have substantial economic impacts on this economy. First, it reduces production of cash and food crops. One result of this production decline is a decrease in export revenues and at the same time an increase in demand for imported food. If foreign borrowing possibilities are limited, the currency must devalue in order to conserve foreign exchange. The devaluation raises the local currency price of imported food and manufactures. If an import competing domestic manufacturing sector exists, production of domestic manufactures will tend to be stimulated by the increased cost of imports. At the same time, demand for domestic manufactures will tend to be weakened by the drought-induced overall decline in income. Income of some agricultural sector participants (those that are large net sellers of food crops), however, might have actually increased due to the drought and devaluation induced increases in the price of food crops. While consumption in many households could be expected to decline, consumption in these net gainer agricultural households might be expected to expand. If consumption baskets differ between households, these effects can be expected to alter the composition of final demand. These are just a few of the many interactions that are triggered by an economic shock such as a drought.

The CGE framework allows one to consider these complex interactions. In many instances, results are driven primarily by the structure of the economy. In other words, the relative size of productive sectors, the nature of household consumption, the importance of international trade, and other relatively well-known factors essentially determine the results. This often robust link between results, data, and non-controversial accounting relationships, with a plausible overlay of assumptions about the behavior of agents, can provide a firm basis for analyzing the overall and distributional implications of large shocks.

A CGE model of Mozambique was used to assess the economic impacts of three climate related experiments (Arndt and Bacou, 2000). The first experiment simulates an unanticipated drought. In this case, farmers adopt an agricultural production plan that cannot be changed based on realized climate outcomes. In the second experiment, farmers alone (i.e., no other agents), are assumed to receive a perfect forecast and respond to it. In the third experiment, both farmers and marketing system agents receive and respond to a perfect forecast.

Drought is simulated through reductions in a technology parameter. Under drought conditions, we assume that the same amount of inputs (labor, seed, etc.) result in less output. The exact technology declines assigned to four groups of agricultural activities in the model are: grain and other export crops - 0.67; basic food crops (primarily vegetables) - 0.75; cassava, raw cashew, raw cotton, and livestock - 0.85; and forestry - 1.00. From a value-of-forecast perspective, the crucial element is not so much the average productivity decline in agriculture as the dispersion across activities. This dispersion allows the possibility to reallocate resources from drought intolerant to more drought tolerant activities (price changes will also influence cropping choice, as discussed below).

With respect to the marketing system (Experiment 3), the following responses might be expected of marketing agents under a credible drought forecast: grain stocks are maintained, ports prepare for increased grain import volumes, the transport systems prepares to distribute inland to normally surplus regions, input suppliers reduce sales expectations, and agricultural processors seek alternative sources of supply. These forecast reactions, as well as the risk reductions inherent in a reliable forecast, are assumed to yield a 2% gain in marketing system efficiency. These simple experiments are designed to illustrate where the largest benefits of forecast information are likely to reside.

The implications of drought for household welfare across the three experiments are shown (Table 8.1). The welfare measure employed is an economic one that focuses on the consumption of commodities (food, manufactures, and services). Greater consumption implies greater welfare. So, the - 4.9% for urban households under unanticipated drought indicates that urban welfare (consumption) has declined. Specifically, the unanticipated drought is equivalent, as far as the urban household is concerned, to the base scenario with 4.9% less income.

As one might expect, rural households bear the brunt of the burden of a drought. More surprisingly, gains to households from even perfect crop reallocation (Experiment 2) are small and skewed towards urban households. In contrast, gains from marketing sector improvements (Experiment 3) are much more significant and more evenly distributed. While urban households reap more than 60% of the value of the (small) gains from crop reallocation under Experiment 2, urban and rural households evenly share the (larger) incremental gains from marketing sector improvements under Experiment 3.

Table 8.1 Household welfare. Equivalent variation on consumption. Percentage change from base

	Base run	Exp. 1	Exp. 2	Exp. 3
Urban	0.0	-4.9	-4.6	-4.0
Rural	0.0	-10.9	-10.7	-10.1
Total	0.0	-7-7	-7.4	-6.9

The large size of marketing margins and the relatively limited scope for producer reactions explain the relative magnitudes of the total benefits. However, some caveats are in order. The true potential for efficiency gains in the marketing system is not really known. These efficiency gains are explored in more detail in the next section. Also, the scope for farmer reaction may be more pronounced when less aggregated data are considered. For example, if labor migrated to the North in anticipation of a forecasted drought in the South, the gains on the production side might be substantially more than the relatively meager gains illustrated in Experiment 2.

Price changes and differential consumption patterns underpin the distributional effects. These results are likely to be more robust than the relative magnitude of the results. The producer's agricultural terms of trade is defined as the producer price of agricultural goods, relative to the producer price of non-agricultural goods. As net sellers of agricultural goods, any increases in the agricultural terms of trade benefit rural households, the vast majority of whom are farmers, since the products they sell can be used to purchase more non-agricultural goods. In Experiment 1, agricultural terms of trade on a producer price basis increase by more than 28%. With drought forecast information, farmers consider both drought tolerance and price in their resource allocation decisions. Farmers will allocate resources to crops whose prices are expected to be high. This increase in the supply of crops that would otherwise have particularly high prices in turn reduces the improvement in agricultural terms of trade to 24%. These relative price declines in agriculture imply that rural households forfeit a large part of the benefits of crop reallocation based on forecast information to urban groups.

In contrast, when efficiencies gained from reactions of marketing sector agents are included (Experiment 3), agricultural terms of trade improve by 24.7% relative to the base, implying a small improvement relative to pure crop reallocation (Experiment 2). This small gain in terms of trade favors rural households. In addition, since marketing costs are particularly important for agricultural commodities, and agricultural commodities are particularly important to rural households (both as income sources and as part of their consumption basket), rural households gain from marketing sector improvements.

Although tentative, these results, in combination with the results reviewed in the literature, point to the benefits of focusing on the marketing sector as targets of seasonal climate forecasts. A rapid appraisal of the Mozambican marketing sector is presented in the following section.

Rapid Appraisal of the Marketing Sector in Mozambique

In order to examine the possibility of reducing domestic marketing margins in the Mozambican economy, a rapid appraisal survey of 13 companies and six public sector institutions was conducted. The survey focused on large companies and institutions that participate in production, transportation, commercialization, and financing of activities involving agricultural goods and food. Due to the diversity of companies surveyed, we prioritized the collection of qualitative information on the main commercialization cost categories. The survey took place in Maputo, Matola, and Beira. Companies and institutions also provided information on how climate variations may affect their activities.

The survey took the form of direct interviews with managers of large companies, including five food processing manufacturers, two food wholesalers, two regional railways branches, one port, one freight company, the largest commercial bank, one international consulting company, and six public institutions, one of which deals with cereal distribution and another of which provides weather forecasting services.

The managers that were interviewed identified main marketing cost categories, presented main factors that drive up those costs, and suggested measures to reduce costs.

The survey was carried out shortly after the massive flooding of February 2000. Most of the firms were facing particularly high transportation costs due to the disruption of National Road 1, which is the main road connecting the North to the South. The floods seriously damaged the road, severing the connection between Southern and Central Mozambique. Firms were forced to switch from road transport to maritime transport to transfer goods between Maputo (in the south) and Beira (in the center).

Since transportation difficulties were a key concern at the time of the survey, inefficiencies in coastal shipping operations, as well as the high cost of these operations, emerged as highly significant in the responses of managers. For example, managers pointed out that shipping a container between Maputo and Beira, which normally takes two days, was about 50% more expensive than the published shipping cost between Asia and Mozambique, which takes about 15 days.

It is not surprising that, given the surge in demand for coastal shipping services caused by the (flood induced) severing of road transport links, prices for coastal shipping were high. Nevertheless, three lessons emerge from this experience. First, purchasers of transport services do have the capacity to substitute between road and maritime transport services, and they exercise that capacity when necessary. Second, having two available systems enhances stability in the face of climate variability. While almost all of the companies interviewed complained about inefficiencies and high costs in maritime transport, the existence of maritime services likely presented an enormous efficiency gain. If maritime service had not existed, the economic disruptions associated with the flood would have been much more substantial.[2] Third, the experience highlights a potential role for climate forecasts. Flooding is a regular event in most provinces in Mozambique, with significant impacts on transport systems. Information on likely rainfall levels and the likely severity of flooding could influence transport decisions both in terms of timing and mode.

A second frequently mentioned factor driving up marketing costs was high costs associated with bank lending. The Commercial Bank of Mozambique (BCM), the country's major bank, identified a number of factors that drive up the cost of credit. Prominent among these is a high degree of risk in making bank loans available. Risk is also a strong consideration for trading and storage activities. This is another area for the potential contribution of climate forecasts. Reliable forecasts would diminish the overall level of risk associated with operations in climate dependent sectors. Since climate dependent sectors represent a large share of GDP (value added from primary agriculture, processed food, and associated marketing activities exceeds 50% of GDP), the overall effects of increased predictability could help to substantially lower risk premiums on an economy-wide basis.

The remaining factors that emerged from the rapid appraisal survey focused on government policy in various arenas. These factors are presumably not amenable to influence by climate forecasts, but nevertheless appear to constrain responses in the marketing sector.

Conclusions

Previous research has considered three levels of analysis of the economic implications of climate forecasts. These levels are 1) farmer responses, 2) responses by participants in the marketing system, and 3) the economy-wide level responses, where the cumulative effect of the various reactions is considered. In this chapter, evidence shows that reactions by marketing agents to climate forecasts may yield the largest economic benefits. This is true in both developed and developing economies. While farmers, at the first level, can adjust their crop mix decisions according to climate forecasts, the scope for reaction, in particular for smallholder producers, may be severely circumscribed. Moreover, an economy-wide analysis that has been undertaken for Mozambique indicates that even if farmers do react, the benefits of these reactions may not accrue primarily to them, but instead to urban consumers.

Even small gains in the efficiency of the marketing system due to climate forecasts could substantially improve welfare. Given these results, a rapid appraisal of marketing margins was used to extract qualitative information from agents operating in the distribution chain. These agents obtain climate information from a variety of sources. The survey indicated two significant areas of marketing costs where forecasts might play a role, which include the timing and mode of transportation and risk reduction associated with credit. Climate information is also an obvious factor that must be considered when planning storage decisions. The conclusions point to a need for more formal analyses of marketing-sector participants in southern African countries. Initially, it is recommended that these efforts should be targeted at areas where climate information can play a fairly significant role, such as storage, transport, and input supply agencies.

Notes

1   Primary sector includes agriculture, fisheries and mining. The secondary sector includes manufacturing, electricity and water, and construction. The tertiary sector comprises services, which includes transport, commerce, banking and finance.
2   The point is not to invalidate concerns from market actors about inefficiencies and high costs associated with coastal shipping. Given the geographical shape, contours of existing land transport infrastructure (road and rail), and quality of natural harbors, coastal shipping appears to be strikingly underutilized.

References

Adams, R., Chen, C., McCarl, B., and Weiher, R (1998), 'The Economic Consequences of ENSO Events: The 1997-98 El Niño and the 1998-99 La Niña', Faculty Paper Series no. 99-2, Texas A&M University.
Arndt, C. and Bacou, M. (2000), 'Economy Wide Effects of Climate Variability and Climate Prediction in Mozambique', *American Journal of Agricultural Economics,* vol. 82, pp. 750-754.
Arndt, C., Hazell, P., and Robinson, S. (2000), 'Economic Value of Climate Forecasts for Agricultural Systems in Africa', in M.V.K. Sivakumar (ed.), *Climate Prediction and Agriculture*, START, Washington, DC.
Arndt, C., Jensen, H.T., and Tarp, F, (2000), 'Stabilization and Structural Adjustment in Mozambique', *Journal of International Development,* vol. 12, pp. 299-323.
Arndt, C., Jensen, H.T., Robinson, S., and Tarp, F. (2000), 'Agricultural Technology and Marketing Margins in Mozambique', *Journal of Development Studies,* vol. 37, pp. 121-137.
Benson, C., and Clay, E. (1998), 'The Impact of Drought on Sub-Saharan African Economies - A Preliminary Examination' World Bank technical paper no. 401, World Bank, Washington, DC.

Council of Economic Advisors (2001), *Economic Report of the President for 1975-1996*, Executive Office of the President, United States Government Printing Office, Washington DC.

Ferris, J.N. (1999), 'Forecasting Global Crop Yields', Department of Agricultural Economics, Michigan State University, Poster Presentation at a Conference of International Association of Agricultural Economists, Berlin, Germany, August 2000.

Francois, J. F. and Reinert, K.A. (eds) (1997), *Applied Methods For Trade Policy Analysis: A Handbook*, Cambridge University Press, New York.

Gommes, R. A. and Petrassi, F. (1994), 'Rainfall Variability and Drought in Sub-Saharan Africa since 1960', Working Paper, FAO Agrometeorological Series no. 9, Rome.

Harrison, M.S.J. and Graham, N.E. (1998), 'Forecast Quality, Forecast Applications and Forecast Value: Cases from Southern African Seasonal Forecasts', United Kingdom Meteorological Office, unpublished.

Ker, A.P. and McGowan, P.J. (1999), 'Weather-Based Adverse Selection and the U.S. Crop Insurance Program - The Private Insurance Company Perspective', Department of Agricultural and Resources Economics, University of Arizona, Tucson.

Macroeconomic Reforms and Regional Integration in Southern Africa (MERISSA). (1998), 'Mozambique Country Study - Agricultural Technology Component', March 1998, MERISSA.

McNew, K. (1997), 'Valuing El Niño Weather Forecasts in Storable Commodity Markets: The Case of U.S. Corn', Department of Agricultural and Resource Economics, University of Maryland.

Ministry of Education (MINED). (1986), *Atlas Geografico*, Ministry of Education, Maputo.

Mjelde, J.W., Peel, D.S., Sonka, S.T., and Lamb, P.J. (1993), 'Characteristics of Climate Forecasts Quality: Implications for Economic Value to Midwestern Corn Producers', *Journal of Climate*, vol. 6, pp. 2175-2187.

National Institute of Statistics (NIS). (1997), 'Electronic Data', Mozambique National Institute of Statistics, Maputo.

National Institute of Statistics (NIS). (1998), *Anuario Estatistico de 1998*, (Statistical Yearbook for 1998), Mozambique National Institute of Statistics, Maputo.

National Institute of Statistics (NIS). (1999), *Results for 1997 Population Census in CD-ROM*, Mozambique National Institute of Statistics, Maputo.

National Institute of Statistics (NIS). (2000), *Electronic Data*, Mozambique National Institute of Statistics, Maputo.

Solow, A.R, Adams, R.F., Bryant, K.J., Legler, D.M., O'Brien, J.J., McCarl, B.A., Nayda, W. and Weiher, R. (1998), 'The Value of Improved ENSO Prediction to U.S. Agriculture', *Climatic Change,* vol. 39, pp. 47-60.

Tibana, R. (1994), 'Mozambique Commodity and Policy Shocks: Terms of Trade Changes, the Socialist 'Big-Push', and the Response of the Economy (1975–86)', Centre for the Study of African Economies Working Paper. Department of Economics, Oxford University.

PART III:
IDENTIFYING USER NEEDS

9 Integrating Indigenous Knowledge and Climate Forecasts in Tanzania

NGANGA KIHUPI, ROBERT KINGAMKONO, HAMISI DIHENGA,
MARGARET KINGAMKONO, AND WINIFRIDA RWAMUGIRA

> ... indigenous forecasting is a tradition to us but foreign to other
> people who control information. Likewise, contemporary or
> scientific climate forecasting is foreign to the people here in
> Goromonzi, but we need to understand the philosophy behind it
> in order to appreciate it at the end. Do not mystify things. We all
> need rainfall information (whether or not is going to rain), but
> precision and timeliness must be the norm.
>
> (Shumba, 1999, p. 64)

Introduction

The climate of 1997/98 attracted worldwide attention, not only because of
extreme weather events, but also because the climate anomalies that caused
many of them were predicted months in advance. In early 1997, scientists
observed that sea surface temperatures in the equatorial Pacific Ocean were
rising sharply over an expanding area, which indicated the development of
a strong El Niño Southern Oscillation (ENSO) episode. Coupled models of
ocean–atmosphere interactions transformed the data into predictions of
anomalous weather extremes in several parts of the globe, many of which
were confirmed to be true by subsequent events.

These model-derived forecasts are based on statistical relationships
between sea surface temperatures and rainfall. Such forecasts do not,
however, represent a unique attempt to predict the seasonal climate. Over

the years, peasant farmers and pastoralists alike have developed their own climate prediction schemes based on observations of the behavior of surrounding nature (Lúcio, 1999; Ngugi, 1999; Shumba, 1999). These schemes have enabled farmers to cope with climate variability and develop numerous adaptation strategies.

Just as sea-surface temperatures in the Indian Ocean can be used as indicators to predict climate patterns over large areas of southern Africa, it may be possible to include local indicators based on indigenous knowledge into a model to provide higher resolution predictions that are more useful to farmers. Although many of the indigenous indicators are not quantifiable and thus cannot be used directly as predictors, they nevertheless often result from changes in atmospheric conditions that can be quantified. If factors such as temperature, humidity, and wind speed are responsible for the changes observed or perceived by local people, then meteorological data from individual stations can be summarized into monthly mean values and correlated with a derived index to identify significant local predictors of the seasonal climate.

This chapter examines the impacts of El Niño rains in selected districts of Tanzania, and briefly appraises the role that seasonal climate forecasts played in 1997/98. Correlations made between a derived aridity index (AI) and the July, August, September, and October mean values of different meteorological parameters show that a number of local indicators may be useful for predicting the climate. The chapter then presents results from a study on indigenous knowledge and climate, and considers how this knowledge is used by farmers.

El Niño in Tanzania: Dissemination and Use of Seasonal Forecasts

Like several countries in the SADC region, Tanzania is heavily dependent on small-scale agriculture to feed its population. The United Republic of Tanzania is located on the south-eastern coast of Africa and is one of the poorest countries in the world. The economy is heavily dependent on agriculture which accounts for approximately 58% of GDP. Agricultural production is predominantly subsistence based, with farmers cultivating on average two hectares of cropland. Most of the rural poor live in villages with very poor infrastructure including roads. This makes access to agricultural inputs and markets difficult.

Tanzania has a climate that varies from tropical along the coast to temperate in the highlands. Although the average precipitation is close to 1000 mm per year (937 mm per year), 50% of the country receives less than 750 mm. Rainfall is highly seasonal and most of the country experiences a dry season between May and October. Rainfall in the northeast is bimodal, with peaks in both October/November and April/May. The short rains are known as *vuli* and the long rains are referred to as *masika*. On the coast, rainy seasons occur from March to May and from November to December. Around Lake Victoria, rainfall is distributed throughout the year, peaking between March and May. In the south, dry seasons occur from May/June to September/October (O'Brien et al., 2000).

As was the case with many people in Africa, most Tanzanians were unfamiliar with the term El Niño before 1997. Since then, the term has become associated in many parts of the country with the heavy rains that wreaked havoc during the 1997/98 season. These rains were the heaviest ever recorded in many parts of Tanzania, and resulted in a collapse of the country's infrastructure and devastation of agricultural production. The rains also claimed several hundred human lives. Livestock and crop diseases that were not common appeared in many areas, and this was attributed to the excessive rains.

There were a few reports on the linkages between the "El Niño" phenomenon and rainfall in Tanzania before the onset of the season. On July 21, 1997, the Directorate of Meteorology presented a statement predicting normal to wetter than normal conditions for the northern coast and the north eastern highlands, and slightly drier than normal to near normal-rainfall conditions over the Lake Victoria basin. Reference was made to the El Niño episode that was under development: "it is likely that atmospheric circulation anomalies, consistent with those experienced during warm episode (El Niño) years should develop in the near future" (*Daily News*, July 22, 1997). The statement cautioned that the forecast should be treated as general advice with modest predictive skill.

In the previous season (1996/97), short rains in the bimodal rainfall areas in the north were poor, and the onset of seasonal rainfall in the south was delayed. Both annual and perennial crops were affected by drought. Thus, not surprisingly, the arrival of rains in the coastal areas and Zanzibar in early October 1997 was first met with considerable optimism for the coming season. Farmers were encouraged to "take advantage of these rains and start planting drought resistant and fast maturing food crop varieties"

(*Daily News*, December 10, 1997). From October onwards, however, newspaper reports on floods and other damages became frequent. On October 21, 1997, the Directorate of Meteorology warned that floods would cause destruction to houses in low-lying areas in Dar es Salaam. Reports of people being left homeless or killed, roads disrupted, buildings damaged and crops destroyed became common. A statement issued by the Directorate of Meteorology on December 4, 1997 reported up to 600% of normal rainfall in areas such as Pemba, Tanga and Zanzibar. This condition was attributed to the influence of El Niño, and was expected to prevail until April 1998 (*Daily News*, December 5, 1997).

The magnitude of the event and extent of the damages are well illustrated in Kilosa, which was one of the most affected districts in the country. This district, made up of 9 divisions, 36 wards, and 137 villages, experienced extraordinarily heavy rains during the short rain (*vuli*) period (November to January). In the 1997/98 season a total of 1,151 mm of rain was received, in contrast to the normal of about 401 mm. In contrast, accumulation during the long rains (*masika*), which start in March and end in May, was only 97 mm above the mean 426 mm. This indicates that the effects of El Niño were more pronounced during the *vuli* season than in *masika* (Figure 9.1).

The heavy rains flooded arable areas in Kilosa, and many farmers were not able to cultivate their fields. Crops that were already planted were washed away, and the topsoil on most good fields was either eroded or covered by sand, gravel, or stones. A substantial number of livestock were killed by floods, and many that survived became malnourished because grazing areas were flooded. It was difficult for farmers to operate machinery or use animal power in the fields because of the excessive rains, and many farmers were unable to fully engage in agricultural operations because their houses were destroyed. Crop yields were reduced further as the result of crop diseases, insect pests, and weeds. Finally, the marketing of crops was hampered, as floods destroyed roads and railway lines. By January, it was anticipated that annual food production would be reduced by 40% (DALDO, 1998).

Scientists disseminated a consensus seasonal climate forecast in September 1997, prior to the onset of the heavy rains. A survey was carried out in 1998 to identify the extent to which seasonal forecasts were received and used by farmers in Tanzania (O'Brien et al., 2000). Of the 299 respondents from Morogoro, Kilosa, Iringa and Mufindi Districts, only

7.4% had received a pre-season forecast. Of these, 82% said they heard the forecasts from the radio,

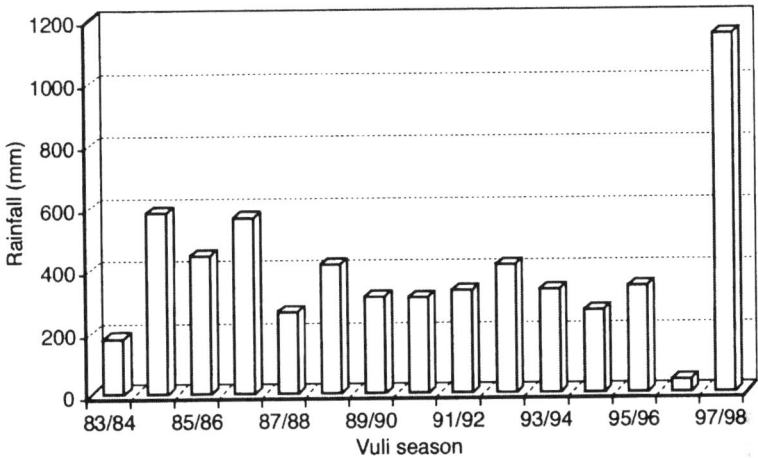

Figure 9.1   *Vuli* rainfall totals at Kilosa for the period 1983/84 – 1997/98

while the rest cited sources such as neighbors, village meetings, newspapers and agricultural extension officers. About 80% of those who heard the forecasts said that they were able to change their planting strategies by planting early or by planting on hilly lands to avoid water logging. Other strategies included planting water tolerant crops, expanding farm size, and storing emergency food. Those who did not change strategies based on the forecast said that the information was not relevant, too general, or not reliable. Some felt that, despite the forecast, there was nothing that they could do.

Seasonal Forecasts: Search for Predictors

The approach used by the national meteorological services is a statistical one which consists of finding the predictors which best explain the predictand(s) by correlation analysis, then choosing predictors with significant correlation to be used for production of models to produce monthly rainfall for the respective seasons. The global sea surface temperatures (SSTs) are most commonly used as predictors, while standardized rainfall indices are used as predictands. Seasonal climate forecasts provide probabilistic estimates of total rainfall relative to a 30-year period. The resolution of the forecasts is rather coarse; one set of terciles can correspond to a region covering several hundred square kilometres. The forecasts provide probabilistic information about total rainfall, but say little about the onset or cessation, the anticipated length of the growing season, or the spatial distribution at a scale that is meaningful to farmers.

The most important questions about rainfall from the farmer's point of view are concerned with the start, end and length of the rainy season including the risk of dry spells within the growing season. Based on an analysis of historical daily weather data, probabilistic profiles of agriculturally important rainfall characteristics can be established. This can lead to the determination of the probable start and end of the rainy season and other characteristics. Combined with climate forecast information, the farmer is able to draw an objective plan, be it a choice of relevant crops to be grown, scheduling of agricultural operations, need for supplemental irrigation, or a complete change of the agricultural system.

The earliest possible onset date refers to a mean date on which rains are expected to start in an area. The earliest possible cessation date refers to the mean date when rains are expected to end. Different approaches have been used to derive such dates. The method adopted by Kingamkono (1993) involves the use of mean daily rainfall data from individual meteorological stations. Other researchers have used grouped data to derive the earliest possible start and end dates of the rainy season (Ilesanmi, 1972; Alusa and Mushi, 1974; Kassase, 1992).

The start and end of the growing season is also of interest to farmers. There are many different definitions of what constitutes a growing season. Often, the start and end of the rains are used to loosely define the growing season. However, a more formal definition considers the growing season in terms of the probability of obtaining a favourable period for non-irrigated crops between two given dates of the year. This refers to the period when soil water, resulting mainly from rain, is freely available to the crop. In the tropics, it is difficult to define a single event to mark the start and end dates of the growing season, as a result of the intermittent and patchy nature of tropical rainfall. A definition should therefore be flexible and designed for the circumstances of the weather system, soil, and crop type. Some approaches make use of the soil water balance to characterize the growing period (Frere and Popov, 1979; Stern, Dennett, and Dale, 1982).

The occurrence of dry spells is of particular interest to farmers in southern Africa. A dry spell does not necessarily refer to a period without rainfall, but can denote a period over which rainfall stays below a certain limit, such as 1.0 mm. There are various methods for predicting the occurrence of dry spells, all based on the statistical analysis of past rainfall events. The simplest methods identify the occurrence of dry spells of a certain length over a certain period at the start of the rainy season, and statistically assess the chance that such a dry spell will materialize. Nevertheless, the frequency of dry spells alone does not capture critical information, since the damaging effect of a dry spell depends on the rainfall received in prior periods and the soil water balance.

To describe the quality of the growing season more objectively, an Aridity Index (AI) has been developed, based on a modified version of the "Rainfall Distribution Index" (RDI) proposed by Kingamkono et al. (1994). The AI combines the median seasonal rainfall total, the median seasonal number of rainy days, and the median length of the growing season. It can be defined as the product of the ratio of the seasonal rainy days to the length of the growing season and the seasonal rainfall total for the respective year. The higher the index, the better is the growing season. This index was used as a predictand in the search for predictors based on indigenous knowledge in a study by Kihupi et al. (2002). The months of July, August, September, and October were selected for correlation analysis. This also happens to be the period when most traditional indicators occur, according to those interviewed.

Climatic data used for the analysis were obtained from the Tanzania Meteorological Agency (on CLICOM files) for some stations within the study area. The data included daily rainfall for all stations, temperature, relative humidity, and wind speed for a few stations. The longest time series ranged from 1960-1999. The CLICOM files were imported into INSTAT computer package (Stern, 1991). The latter is capable of analysing climatic data including dates for start or end of the growing season using set conditions. The results for Dodoma are presented as an example of the potential integration of local data in climate forecasts. Table 9.1 shows that the August mean temperature, the August and September mean maximum temperature, the September relative humidity and the September mean wind speed had correlation coefficients of greater than or less than 0.3 and thus are potential candidates for inclusion in predictive models.[1] Occurrence of long dry spells at the beginning of the growing season appear to influence the season negatively. These results corroborate observations made by local people.

Table 9.1 Correlation between Aridity Index and predictor for Dodoma

Predictor	Correlation coefficient
July mean temperature	0.289
August mean temperature	0.368
September mean temperature	0.191
October mean temperature	0.173
July mean maximum temperature	0.265
August mean maximum temperature	0.344
September mean maximum temperature	0.332
October mean maximum temperature	0.057
August mean relative humidity	0.015
September mean relative humidity	-0.356
October mean relative humidity	-0.035
August mean wind speed	0.236
September mean wind speed	0.301
October mean wind speed	0.283
Long dry spells at the beginning of the growing season	-0.446
Long dry spells late in the growing season	0.087

Indigenous Knowledge of Weather and Climate

As indicated above, several methods are being explored to make climate information more useful to the end-user, in this case, farmers. To reduce the impacts of events such as the floods of the 1997/98 season, it is necessary to put in place measures to mitigate and help manage such hazards. The use of current "scientific forecasts" and methods to assist in the management of climate variability may, however, be limited in the African context in which several farmers and users find themselves. Forecasts may be more useful if ways are found to integrate local knowledge, which has enabled generations to live through severe floods and droughts, into current management decision-making strategies. Indeed, results from forecast use in the region and elsewhere have shown that in the drive to make forecasts useful and meaningful, scientific knowledge cannot be posited as exclusive, and it is important to be mindful that users operate in multiple cognitive frameworks that need to be understood (Orlove and Tosteson, 1999; Roncoli, Ingram, and Kirshen, 2002). With this in mind, an examination of the use of traditional forecasts was undertaken in Tanzania.

The use of indigenous knowledge for forecasting extreme weather events is well known and is used by most farmers in Southern Africa (Lúcio, 1999; Ngugi, 1999; Shumba, 1999). Although farmers listen to forecasts from radios, most of them (particularly the poor and marginalized) use their traditional knowledge systems of climate forecasting as controls. The more that contemporary climate forecasting information deviates from these controls, the less it is used for planning purposes. The signs vary depending on the location, experience, and the traditional culture of the community. Where traditional culture has been mixed by religion, especially Christianity, such indigenous knowledge is disappearing.

To explore the role of indigenous knowledge in climate prediction, a study was carried out in a total of nine districts selected from three regions in the Central and Northern zones of the country prior to the 2000/2001 season (Kihupi et al., 2002).[2] Over 90% of the 914 respondents surveyed by Kihupi et al. (2002) acknowledge the existence of indigenous indicators or signs used to warn of an impending climate anomaly (drought/floods). The village elders or chairpersons are the main custodians of forecast information based on indigenous indicators. The mode of communication of such information is largely informal, through individual contacts.

Indicators based on the appearance of plants seem to be most important (Table 9.2). Flowering density of certain trees such as the Nandi flame tree (*Delonix regia*), mangoes (*Mangifera indica*), and certain *Accacia spp.* locally known as *Migunga*, immature dropping of fruits by certain tree species, shedding of leaves of the sycamore fig (*Ficus sycomorus*) and exudance of water from the leaves of *Albizia schimperiana* before the onset of the rains are indicative of the type of the season to be expected. For example, in Himiti village (Babati district), mango trees were observed to have flowered more than normal in October, 2000, which was interpreted as a forecast for a good season, as turned out to be the case. Several other tree types in the central zone appeared to have blossomed more than normal around September 2000, which was also taken to indicate good rains. These areas also turned out to have received good rains during the 2000/2001 rainy season.

Ambient temperature is the next most important indicator. High temperatures, especially at night during the months of September to November, are considered to signal good rains and a long growing season in all of the areas visited. Low night temperatures are also said to indicate late onset of the rainy season. Elders in Himiti and Managhat villages in Babati district recount that temperatures prior to the 1999/2000 rainy season were generally low, indicating poor rains, which turned out to be the case for that season. Temperatures prior to the 2000/2001 rainy season were higher and based on this, the elders predicted good rains, which again turned out to be the case.

Other indicators are listed in Table 9.2 in order of importance. From these and other indicators, elders in the villages predicted that the 2000/2001 season would have good rains. The signs, including strong winds, hot sun and mist during the months of August to October, were quite similar but not as intense as those observed in 1997, just prior to the El Niño rains. It would appear that the local communities are knowledgeable not only of whether there might be floods or drought in the coming season, but also on whether the season is going to be long or short and evenly or poorly distributed rains with an early or late onset.

The accuracy of recollection by the local people affects the way indigenous knowledge has been built up and passed on over generations. Certain episodes are recalled vividly. For instance, the 1942 famine as recalled by the Himiti village farmers in Babati district was a result of a serious drought. They termed it "Njaa ya bakuli," literally "bowl hunger," because they were provided with relief food, rationed out in bowls, of maize flour, beans and cooking oil. In 1954 another serious drought was termed by the Himiti villagers as "Njaa ya makopa," meaning dried cassava, the relief food provided by the government. Both 1942 and 1954 were La Niña years.

Table 9.2 Traditional indicators used in Central and Northern Tanzania

Indicator	Indication	Time of occurrence	Response (%)
Appearance of plants	Higher than normal flowering density of Nandi flame tree (*Delonix regia*), mangoes (*Mangifera indica*), *Commiphora africana*, ("Mpome"), *Adansonia digitata* ("Mbuyu"), *Kigelia africana*, tree with white flowers locally known as "Ormukutan" (Arusha) and certain *Accacia spp.* (Migunga") – good rains	Sept./ Nov.	25.9
	Immature dropping of fruits – drought	Sept./Oct.	
	Shedding of leaves, flowering and bearing fruits fast e.g., *Ficus sycomorus* – good rains	Aug./Sept.	
	Water drips from leaves of *Albizia schimperiana* and "Mkalakala" (Swahili) – good rains	Sept./Oct.	
Ambient temp- eratures	Higher than normal especially at night – good rains	Aug.-Oct.	14.3
	Lower than normal – poor rains		
Behaviour of domestic animals	Increased libido in goats with frequent matings – good rains	Aug./Sept.	9.7
	Increased libido in donkeys – poor rains	Aug.	
	Presence of numerous watery cysts in goats' guts – good rains	Aug./Sept.	
	Cattle running happily on their way back from grazing in the evening and reluctant to go for grazing the next morning – good rains	Aug./Sept.	
Appearance of the moon	First rains just before the appearance of the new moon – good rains	Oct./Nov.	7.5
	Full moon covered by clouds – good rains	Oct./Nov.	
	A halo surrounding the moon – good rains	Sept.-Nov.	
Appearance of various insects	Increased numbers of red ants – good rains	Sept./Oct.	7.0
	Occurrence of army worms –drought	Feb.	
	Mushrooming of anthills which are moist – good rains	Aug./Sept.	

Thunder-storms	Heavy thunderstorms – generally good rains	Throughout the season	6.4
Clouds	Heavy dark clouds on mountains – good rains	Aug./Sept.	6.4
Appearance of two bright stars	One star from the eastern direction and the other from the western direction – good rains	August	6.2
Wind direction	Change of direction from W-E to S-N – good rains	Sept./Oct.	5.2
	N-S to E-W – poor rains	Sept./Oct.	
Appearance of birds	Swallows or "Mbayuwayu" in Swahili and "Narmo" (local name) birds – good rains	Oct./Nov.	4.3
	"Demsi" birds – bad season	Sept./Oct.	
Behavior of wild animals	Frequent roaring of lions/hippopotamus – good rains	Aug./Sept.	2.3
	Appearance of Pangolins ("Kakakuona" in Swahili) – good rains	Aug.-Oct.	
Witch-doctor's revelations	Special black stones becoming moist	July-Nov.	1.7
	Dreams	July-Nov.	
Wind swirls	Frequent occurrence of wind swirls- good rains	Sept./Oct.	0.9
Earth-quakes/tremors	Occurrence of weak earthquakes or tremors – good rains	Aug./Sept.	0.9
Water sources	Drying up of water sources e.g., springs, wells, rivers at a faster rate – good rains	Aug./Sept.	0.5
Birth of babies	Birth of many baby girls prior to the onset of the rains – good rains	Aug.-Oct.	0.4
Mists and rumbling sounds in mountains or hills	Rumbling sounds especially when temperatures are high at night and/or when the mountains are covered with "heavy white clouds" locally known as "Gitsimi" – good rains	July-Oct.	0.4

Many elders in the surveyed villages recalled that there were floods in 1962/63, an El Niño year. Unfortunately no warning signs could be recalled by the elders, with the exception of one who recalled a mild earthquake shortly before the rains, which is usually indicative of heavy rains. Similar events were recalled in 1973/74 and 1974/75 (Kihupi et al. 2001). The

1980s are recalled by many as having been relatively good years with very few climatic anomalies.

The most recent episode witnessed by many are the 1997/98 floods. According to the elders in Himiti village, this condition was expected from their own indigenous predictions. They went as far as requesting the "chief rain maker" to mitigate the situation, which he declined to do as this was "God's own wish." This was the famous El Niño episode which brought awareness to most people in the country of the existence of such a phenomenon.

Conclusions

Like the contemporary seasonal climate forecasts, indigenous methods are integrated in the farm-level decision making process. Both methods are thought by the majority to be useful and generally reliable, not only in this study but in several other cases in the region (e.g., Lúcio, 1999; Ngugi, 1999; Shumba, 1999; Phillips, Makaudze, and Unganai, 2001; Patt, 2001; Vogel, 2000). Indigenous forecasting knowledge is to a large extent a reflection of the interactions between the community and the environment that has developed over a long time period. The information has been tried and tested, and has slowly been integrated into the community's culture. It is therefore easier to reject the "foreign" knowledge than the indigenous knowledge that is well trusted. To increase acceptability, the "foreign" knowledge must prove to be as useful as or even better than indigenous knowledge.

Since farmers use a number of different signs to predict climate, it is important to identify and classify the indicators and methods that are specific to a particular locality. So far, no pattern or trend has been identified to indicate which specific signs or methods can be used to predict weather events in different areas or communities. An attempt to facilitate the integration of indigenous and contemporary climate forecasting information challenges conventional rural development paradigms centered around giving the farmer that which is new, and instead builds on what the farmer knows (Ngugi, 1999). There is clearly a need to study and characterize indigenous indicators for the purpose of integrating traditional forecasting methods into more formal seasonal climate forecasting techniques.

Notes

1   Those that had correlation coefficients of greater than or less than 0.3 were considered to be significant. This is the cut-off point that has been used in the pre-season capacity building workshops on seasonal forecasting.

2   Participatory Rural Appraisal (PRA) methods viz. key informant interviews and focus group discussions were used. In addition, questionnaires were also administered to different groups of elders other than those involved in the PRA interviews. A check list that included issues on knowledge on climate, seasonal climate forecasting, traditional indicators, past events with focus on climate anomalies and associated coping strategies, guided the interviews. A group of elders aged 50 years and above was involved in the PRA for each selected village.

References

Alusa, A.L. and Mushi, M.T. (1974), 'A Study of the Onset, Duration, and Cessation of the Rains in East Africa', Preprints, International Tropical Meteorology Meeting, Nairobi, Kenya, pp. 133-140.

District Agricultural and Livestock Development Officer (DALDO) (1998), *Report to the Regional Agricultural and Livestock Development Officer (RALDO)*, Kilosa, January 28, 1998.

Frère, M. and Popov, G.F. (1979), 'Agrometeorological Crop Monitoring and Forecasting', FAO Plant Production and Protection Paper, No. 17, Rome.

Ilesanmi, O.O. (1972), 'An Empirical Formulation of the Onset, Advance and Retreat of Rainfall in Nigeria', *Journal of Tropical Geography*, vol. 34, pp. 17-24.

Kassasse, C.I. (1992), *Determination of Effective Length of Growing Season in Tanzania*. M.Sc. Dissertation, Sokoine University of Agriculture, Morogoro.

Kihupi, N.I., Rwamugira, W., Kingamkono, M., Mhita, M. and O'Brien, K. (2002), *Promotion and Integration of Indigenous Knowledge in Seasonal Climate Forecasts*. Report to Drought Monitoring Centre, Harare, Zimbabwe, NOAA/OPG and USAID/OFDA.

Kingamkono, R.M.L. (1993), *The Effective Length of Growing Season in Tanzania*. M.Sc. Dissertation, Sokoine University of Agriculture, Morogoro.

Kingamkono, R.M.L., Kihupi, N.I., and Dihenga, H.O. (1994), 'Length of Growing Season vis-à-vis Rainfall Distribution in Tanzania', In C. Lungo (ed.), *Proceedings of the Fifth Annual Scientific Conference of the SADC – Land and Water Management Research Programme*, Oct. 10-14, 1994, Harare, Zimbabwe, pp. 141-155.

Lúcio, F.D.F. (1999), Use of Contemporary and Indigenous Climate Forecast Information for Farm Level Decision Making in Mozambique, Consultancy report, UNDP/UNSO.

Ngugi, R.K. (1999), *Use of Indigenous and Contemporary Knowledge on Climate and Drought Forecasting Information in Mwingi District, Kenya,* Consultancy Report to the United Nations Development Program Office to Combat Desertification and Drought, UNSO/UNDP/WMO.

O'Brien, K., Sygna, L., Naess. L.O., Kingamkono, R., and Hochobeb, B. (2000), *Is Information Enough? User Responses to Seasonal Climate Forecasts in Southern Africa.* CICERO Report 2000:3, Oslo, Norway.

Orlove, B. and Tosteson, J. (1999), 'The Application of Seasonal to Interannual Climate Forecasts based on El Niño – Southern Oscillation (ENSO) Events: Lessons Learned from Australia, Brazil, Ethiopia, Peru and Zimbabwe', Working papers in Environmental Policy, Institute of International Studies, University of California, Berkeley.

Patt, A. (2001), 'Understanding Uncertainty: Forecasting Seasonal Climate for Farmers in Zimbabwe', *Risk Decision and Policy*, vol. 6, no. 2, pp. 1-15.

Phillips, J.G., Makaudze, E. and Unganai, L. (2001), 'Current and Potential Use of Climate Forecasts for Resource-poor Farmers in Zimbabwe', in American Society of Agronomy Special Publication #63. *Impacts of El Niño and Climate Variability on Agriculture*, pp. 87-100.

Roncoli, C., Ingram, K. and Kirshen, P. (2002), 'Reading the Rains: Local Knowledge and Rainfall Forecasting in Burkina Faso', *Society and Natural Resources,* vol. 15, pp. 411-430.

Shumba, O. (1999), Coping with Drought: Status of Integrating Contemporary and Indigenous Climate/Drought Forecasting in Communal Areas of Zimbabwe, Consultancy Report to the United Nations Development Program Office to Combat Dessertification and Drought, UNSO/UNDP/WMO.

Stern, R.D., Dennett, M.D., and Dale, I.C. (1982), 'Analysis of Daily Rainfall Measurements to give Agronomically Useful Results. I. Direct Methods', *Experimental Agriculture*, vol. 18, pp. 233-236.

Stern, R. (1991), *INSTAT Climatic Guide*. Statistical Services Centre, University of Reading, UK.

Vogel, C.H. (2000), 'Usable Science: An Assessment of Long-term Seasonal Forecasts amongst Farmers in Rural Areas of South Africa', *South African Geographical Journal*, vol. 82, pp. 107-116.

10 Forecasts and Commercial Agriculture: A Survey of User Needs in South Africa

EMSIE KLOPPER AND ANNA BARTMAN

Introduction

Climate is one of many factors that farmers consider when making production-related decisions. Reliable predictions of expected climate conditions can counter uncertainty and contribute to improved decision-making. The efficient application of forecasts depends on the nature and context in which users make decisions, and on various characteristics of the information on which the decisions are based. Although seasonal climate forecasts are relevant to a number of groups in South Africa, farmers represent the largest single group of potential users. In an agricultural environment, seasonal climate forecasts could contribute valuable information for investment decisions about farm buildings, field machinery, crop drying, and irrigation equipment, and for managing flood and drought risks.

Limited accuracy and skill in predicting extreme events are amongst the most common impediments to the use of climate predictions in business, operations, and planning decisions (Easterling, 1986; Sonka et al., 1988). Other limiting factors include reliability, spatial resolution, and timeliness of forecasts. Given these impediments, meteorological services are increasingly forced to consider the nature of their products, the value to end-users, and the user profiles of their clients. Privatization of the South African Weather Service[1] in 2001 has necessitated a careful assessment of the products offered, a better understanding of their target groups, and improvements in the delivery of services. Feedback from users is thus an important prerequisite for tailoring climate information to meet specific needs, for distributing information in a timely manner, and for providing support services.

This chapter assesses the value of forecasts issued by the South African Weather Service to commercial farmers, as well as possibilities for increasing the value through an expanded product line and improved dissemination. The uptake, use, and information requirements of commercial farmers in South Africa were analyzed through questionnaires sent out after the 1997/98, 1998/99 and 1999/2000 summer rainfall seasons. Results from these surveys show that forecasts are indeed used by those who sought out the information, but that their value would be greater if they were tailored and packaged to meet the specific needs of different farmers.

Agriculture and Climate Variability in South Africa

The Republic of South Africa covers an area of 122.3 million hectares, of which 84% is used for agricultural purposes. Agricultural land is mainly used for grazing, but about 13% of the area can be used for crop production. High potential arable land comprises only 22% of the total arable land. Slightly more than 1.2 million hectares are under irrigation (National Department of Agriculture, 1997).

Farmers are a diverse group in South Africa, as a result of the widely differing climatic conditions of the country, a wide range of agricultural products, and the differential socio-economic history and status of individual farmers. Small-scale farmers are often economically and technically more constrained in their ability to make use of seasonal forecasts, in contrast to commercial farmers who can capitalize on forecast information due to wider access to information and infrastructure (e.g., credit and other inputs) (Vogel, 2000). Maize, wheat, and sugar cane are the major consumer products traditionally produced by commercial farmers. Recently, efforts have been made by the government to support the emergence of a more diverse structure of agricultural production. This is expected to increase the number of successful smallholding enterprises.

The most important factor limiting agricultural production in the country is the availability of water. Already, almost 50% of South Africa's water is used for agricultural purposes (National Department of

Agriculture, 1997). Rainfall is highly seasonal and distributed unevenly across the country, with humid, subtropical conditions occurring in the east and dry, desert conditions in the west. More than 80% of the annual rainfall occurs between October and March (Taljaard, 1986; Tyson, 1986).

On a seasonal time-scale, the El Niño Southern Oscillation (ENSO) phenomenon contributes to about 30% of the rainfall variability in South Africa (Lindesay, 1988). In general, the country tends to experience dry conditions during El Niño events, and normal to wet conditions during La Niña events (Lindesay, 1988; van Heerden, Terblanche, and Schulze, 1988; Schulze, 1989). The ability to provide advanced notice of such phenomena has greatly increased over recent years for the region (see Mason et al., 1996; Mason, 1998). While it is by no means a "perfect" science, advance warnings of a potential event has much to offer in mitigating the impact of drought and improving food security in drought-prone countries, including South Africa (Glantz, Betsill, and Crandall, 1997).

Although the 1997/98 El Niño was the strongest on record, not all of South Africa received below-normal rainfall (Landman and Mason, 1999a, 1999b). Some regions had adequate rainfall due to incursions of moist air from the Indian Ocean. The 1998/99 season was characterized by La Niña conditions, but a stationary area of tropical convergence developed in the Mozambique channel and caused dry, hot weather over the country due to subsidence. The 1999/2000 La Niña season will be long remembered for the tropical cyclone Eline, which caused high rainfall and floods over the northeastern parts of South Africa and its neighboring countries during February 2000. This contributed to above-normal seasonal rainfall over these areas. With this background on the progression and impact of the ENSO phenomenon during the late 1990s, we now examine how seasonal forecast information was perceived and used.

Assessing Forecast Value

Interest in assessing the value of forecasts of all types has increased in recent years. This interest is "due in part to efforts by weather services to extend their forecast services and thereby enhance their revenues, as well as efforts by users to improve the efficiency of their activities" (Murphy, 1994, p. 69). It is widely recognized that the value of climate information greatly exceeds the cost of producing a seasonal climate outlook (Johnson and Holt, 1997). Yet weather and seasonal forecasts have no intrinsic value in an economic sense. They acquire value by influencing the behavior of individuals and organizations (Murphy, 1994). To have value, climate information needs to be understood and used in the decision-making process.

Forecast value studies represent a potentially important source of information on the nature of how users make decisions, as well as on relationships between the quality of forecasts and their economic value (Murphy, 1994). One of the outcomes of in-depth forecast value studies is that they can provide information about how the producers of forecasts can enhance both the product and its dissemination. Many users desire access to experts, such as weather scientists or agricultural advisers, to help them apply the seasonal climate forecast information. Other users require regionally-specific information tied to specific areas of interest or key production regions of competitors. There is also a demand for long-term forecasts of conditions beyond 90 days, and for a variety of historical climate analyses, recent averages and extremes, and listings of recent years that are analogous to current forecasts.

Agricultural production, like most economic activities, is not simply the result of deciding what to produce. Important management choices have to be made regarding a range of input variables, such as crops and seed variety, fertilization rates, cultivars, seeding rates, applied nitrogen rates and planting dates. Information support systems needed to optimize decisions include expected pasture growth, animal live-weight gains and optimal stocking rates, grain production, optimal fertilizer applications, projected yields, frost prevention measures, and pest and disease management (Kininmonth, 1994). It is thus necessary to acknowledge the broad nature of these various agricultural decisions and the multiple ways that forecasts can influence a farmer's decision-making strategies. The practical value of seasonal forecasts depends, to a large extent, on the decision-makers' willingness and ability to utilize a flexible management style (Sonka, 1986). A user must be able to re-evaluate decisions as new information becomes available and should not "routinize" decision-making (Simon, 1984).

Production and Dissemination of Seasonal Forecast Information

The South African user community receives a wide spectrum of seasonal climate forecasts. The Southern African Regional Climate Outlook Forum (SARCOF), for example, produces a regional seasonal climate outlook for fourteen member countries of the Southern African Development Community (SADC) (Harrison, 1998). The seasonal climate outlooks generated through the SARCOF process are then disseminated to user communities in South Africa.

At the national scale, a seasonal climate forecast for South Africa is compiled on a monthly basis by the South Africa Weather Services. Model output provided by international institutions such as International Research Institute for Climate Prediction (IRI) is combined with regional models to produce a single consolidated seasonal outlook that is distributed to various users in South Africa (Mason, 1998; Klopper, Landman, and van Heerden, 1998; Landman and Mason, 1999a; Tennant, 1999). The seasonal outlook consists of a 3-month mean temperature and accumulated rainfall forecast for one to two seasons in advance. Information on the current and expected state of ocean and atmospheric parameters influencing Southern Africa are also included in the forecasts.

Both the regional and national seasonal forecast products are disseminated to users through the Long-term Operational Group Information Centre (LOGIC), an affiliate of the South African Weather Service. LOGIC was established in 1997 as a result of the large number of enquiries on seasonal forecasts, triggered by the approaching ENSO. The aim of LOGIC is to handle enquiries from end-users concerning long-term forecasts. Different modes of communication are used to serve the end-user. Besides answering telephone enquiries, LOGIC issues information through a fax-on-demand system, mail, e-mail and the Internet. Many news interviews and television presentations are also conducted using information supplied through LOGIC. All relevant information, including the latest monthly and seasonal outlooks and observations, are displayed in the office so that real-time information is available to serve the end-user. In addition to the 3-month rolling forecasts, LOGIC issues rolling forecasts (updated monthly) for different seasonal lead times, ranging between 2 and 6 months.

The growing demand for seasonal climate information has been noteworthy. Nine months prior to the 1997/98 rainfall season in South Africa, warnings of a developing El Niño event were issued to the public.

Great interest in the forecasted event led to more than 2000 telephone enquiries to the LOGIC office at the Weather Service between the period between August 1997 and March 1998 (Klopper, 1999). In turn, the LOGIC office sent out between 400 and 500 faxes, letters, and e-mails to disseminate seasonal outlooks. It is clear that users both needed and were willing to ask for information.

The Use of Forecasts by the Commercial Agricultural Sector

To discover whether and how the forecasts information was being put to use by those who requested it, questionnaires were compiled and sent to end-users during the three summer rainfall seasons from 1997 to 2000. The main objective of the questionnaires was to determine whether the seasonal outlook products reached the end-user community in time, and to assess how they incorporated the information into their decision-making (Klopper, 1999). The end-users that were selected for the survey included subscribers to the seasonal outlook products in South Africa (from the LOGIC database). The majority of the responses were obtained from the agricultural sector, with commercial farmers representing 75% of the respondents in 1997/98, 82% in 1998/99, and 77% in 1999/2000. The main activities of these agricultural end-users include both crop production (maize, wheat, sorghum, etc.) and livestock enterprises.

Not surprisingly, the respondents relied primarily on LOGIC as a source of information, followed by the media. This suggests that effective interaction and cooperation between the producers of forecasts and the media is important. Differing interpretations by journalists or presenters can create an additional complication to forecast dissemination, underscoring the need for clear explanations of each forecast, its limitations, and use.

During the 1997/98 season, 79% of the commercial farmers that responded to the questionnaire modified their production decisions as a result of the seasonal forecast of an El Niño event. In response to the La Niña forecasts of 1998/99 and 1999/2000, more than 80% of respondents changed their production decisions. Decisions that were modified included, among other things, planting schedules and farm inputs (e.g., fertilizer amounts and dates) (Table 10.1).

Table 10.1 Summary of modifications of decisions by commercial farmers during the 1997/98 and 1998/99 summer rainfall seasons

	El Niño (1997/98)	La Niña (1998/99 and 1999/2000)
Crop farmers	- Changed crop varieties (selection of short season maize hybrids, introduction of faster growing varieties)	- Changed crop varieties - Planted crops on well-drained soils
	- Planted more hay and fodder (reduced amount of maize to keep enough irrigation pastures for the stock)	- Sprayed and planted earlier - Increased plant population - Increased fertilizer applications
	- Delayed or earlier planting (not planted at all in specific cases)	- Changed irrigation schedules
	- Reduced seeding	- Changed land preparation
	- Conserved moisture (water)	- Changed harvest programs
	- Used fertilizers conservatively	
Live-stock farmers	- Reduced stock (sold cattle before the season started; sold 30% of cattle)	- Maintained or increased livestock numbers
	- Developed additional trough systems for stock	

Decisions of a more general nature were also reported. For example, many farmers planned more conservatively and improved water management as a result of the El Niño forecast. Risk profiles were adjusted, and labor recruitment plans were changed. Some farmers increased cooling systems, or changed temperature of drying equipment to save electricity. Government departments developed a drought policy, and advised clients based on an anticipated dry season. The La Niña forecasts for the following two seasons resulted in an increase in credit availability, and farmers reported taking greater financial risks. Labor requirements were also planned more accurately as a result of the forecasts.

Most commercial farmers (75% of the respondents) encountered little difficulty in interpreting and applying the seasonal outlook information. However, one problem mentioned repeatedly was that the forecasts were not tailored to the specific geographical location of the farm. As echoed elsewhere in this book, it is considered a major limitation that the forecast given is not specific enough to the particular area in which the farmer lives.

While the forecasts had an impact on farmers' decisions, many other factors besides anticipated rainfall influence the planning process. Financial, economic, and demographic factors are crucial to the decision-making process of commercial farmers. Indeed, the most prominent factors are usually related to the current market situation. For example, increased transportation costs resulting from high fuels prices may influence production decisions, as can high labor costs. Other factors that influence decisions include the general availability of water and the quality of other resources. Past experience and knowledge of the environment, though sometimes difficult to capture and quantify, are also critical factors that influence agricultural production decisions. Some of these factors are summarized below:

- Economic factors: The availability of financing, the rate of interest, and the individual farmer's financial position.

- Market conditions: The prices or contracts negotiated for certain merchandise, international trends, the demand for a specific item.

- Water availability: The level of dams, the level of the sub-surface water table, and the probability of run-off from catchment areas during the season.

- Quality of resources: Soil quality, quality and availability of pasture, and condition of the grasslands and other natural vegetation.

Seasonal forecasts are part of a larger, complex suite of issues, which include cultural, socio-economic and other daily realities that influence the uptake of the seasonal outlooks. The wider socio-political and economic situation of the country should be taken in to account when evaluating the uptake and use of seasonal climate forecasts. Changes in the post-1994 South African government policy, for example, have had a particularly strong impact on farmers' decision-making environments. These policy changes include land reforms, deregulation of agricultural boards, and the re-evaluation of drought subsidies.

Improving the Forecasts

From the survey results, it is evident that the seasonal forecast products are both useful and relevant to commercial farmers in South Africa. Many respondents expressed a need for this information, since it does have an impact on agricultural decisions. It is also clear, however, that improved communication between users and producers of seasonal outlook products is fundamental to determining which climate factors are directly relevant to the commercial farmer.

To enhance the application of seasonal forecasts, farmers require additional, tailor-made information such as forecasts of crop maturity dates, regional crop yields, drought and flood risks, cold and heat waves, fire hazard risk, and pest and disease risk information. Additional indices for agriculture would thus aid in the decision-making process. Satellite images of the region showing cloud formation and winds, comparison with historical events, ongoing evaluation as the season progresses, synoptic charts and more accurate climate statistics would also contribute towards optimizing use of forecast products. Farmers also require more area-specific forecasts and explanations of some climate scenarios, such as for an extended drought.

The most critical time for receiving a forecast seems to be just before the start of the summer rainfall season. Respondents require a pre-season forecast that provides an indication of rainfall conditions for the entire season during late August or early September. This corresponds to the time when farmers purchase seed and prepare land. Updates of these forecasts are also required to fine-tune production decisions, such as identifying when fruits can be picked or other crops harvested. The updated forecasts are also important during planting and pricing periods. Furthermore, farmers need updates shortly before excessive rain or cold spells.

The time period for which the seasonal forecasts are made is indeed an important determinant of its value. In prioritizing future climate prediction research needs, Mjelde et al. (1997) found that foreknowledge of the late growing season was of most value to the producer, since the late growing season is the critical time for crop tasselling and grain fill. This indicates that the most important stages of production in terms of value of climate forecasts are stages that require longer lead times. Improvements in late season forecasts have a greater economic value than improvements in early season forecast quality. Lead-time is thus critical: simple rules such as forecasting all events two or three months in advance are unlikely to fully satisfy the decision-maker's needs. Longer lead times are generally more beneficial to farmers, but such information is more uncertain. Long-term forecasts should be accompanied by medium-term updates, since the most significant economic benefits are derived from the latter (Lyakhov, 1994).

Although the 1997/98 El Niño was the strongest observed on record, forecasters advised end-users in South Africa that the impact would be less than during the 1982/83 El Niño season. This was due to the surplus of available water and high soil moisture content that was carried over from the previous good rainfall season. Unfortunately, nearly all media reports associated El Niño with drought, sending out a negative message to the public in anticipation of the 1997/98 rainfall season. The media is an important and powerful instrument that can be used for changing perceptions, creating public awareness, and improving the understanding of risk reduction. It is evident that guidelines are needed to relate information to the media in such a way that it reaches the end-user in a factually based manner.

Commercial farmers are not always aware of the potential use and value of seasonal outlooks. It is the responsibility of the producers of such information to assist farmers in better utilizing and improving information flow. An educational program, including workshops and public awareness information sessions, is needed to educate users about both the existence and the use of these products. Training of extension officers (usually the links between farmers and ministries and producers of information) is, for example, currently underway by the Weather Service and the Agricultural Research Council in Pretoria. Innovative forecast design strategies are also necessary to make end-users more responsive to the information. The value of these products can further be increased by collaboration between sectors, such as agriculture and water management. Thus, the effective use of information requires rigorous research efforts to determine how to tailor that input to the specific needs of decision-makers.

The seasonal climate information products available, and the dissemination methods used, are currently more readily accessible to commercial farmers, but a large majority of the farmers in South Africa are subsistence and emerging farmers. The diverse needs of these users of seasonal climate information products should also continue to be analyzed in order to render an appropriate service in terms of product range, dissemination and education. Various projects are being launched to address this need in South Africa. The strategic position of South Africa and the country's expertise and specialist services enables it to make a large contribution towards agricultural development to all its neighboring states, and Africa as a whole.

While forecasts do have some value and are being used by farmers, it is clear that there is a need for more effective dissemination, communication and education, as well a continuing need for better forecasts. Greater and on-going interaction between users and producers of forecasts can potentially ensure that farmers have an additional tool to assist them in their decision-making process.

Note

1 South African Weather Service was previously called the South African Weather Bureau.

References

Easterling, W.E., (1986), 'Subscribers to the NOAA Monthly and Seasonal Weather Outlook', *Bulletin of the American Meteorological. Society,* vol. 67, pp. 402-408.

Glantz, M., Betsill, M. and Crandall, K. (1997), *Food Security in Southern Africa: Assessing the Use and Value of ENSO Information,* Environmental and Societal Impacts Group, National Centre for Atmospheric Research, Colorado.

Harrison, M.S.J., (1998), 'First Report of the Southern African Regional Climate Outlook Forum', Unpublished report, U.K. Meteorological Office.

Johnson, S.R. and Holt, M.T. (1997), 'The Value of Weather Information', in R.W. Katz and A.H. Murphy (eds), *Economic Value of Weather and Climate Forecasts,* Cambridge University Press, Cambridge.

Kininmonth, W.R., (1994), 'Economic benefits of long-range weather forecasting', *Geneva,* WMO/TD-No.630, pp. 226-229

Klopper, E., (1999), 'The Use of Seasonal Forecasts in South Africa during the 1997/98 Rainfall Season', *Water SA,* vol. 25, no. 3, pp. 311-316.

Klopper, E., Landman, W.A. and van Heerden, J. (1998), 'The predictability of seasonal maximum temperature in South Africa', *International Journal of Climatology*, vol. 18, pp. 741-758.

Landman, W.A. and Mason, S.J., (1999a), 'Operational Long-lead Prediction of South African Rainfall using Canonical Correlation Analysis', *International Journal of Climatology, vol.* 19, pp. 1073-1090.

Landman. W.A. and Mason, S.J. (1999b), 'Change in the Association between Indian Ocean Sea-surface Temperatures and Summer Rainfall over South Africa and Namibia', *Journal of Climatology*, vol. 19, pp. 1477-1492.

Lindesay, J.A. (1988), 'South African Rainfall, the Southern Hemisphere Semi-annual cycle', *International Journal of Climatology*, vol. 8, pp. 17-30.

Lyakhov, A.A. (1994), 'Analysis of Economic Benefits of Short Term, Medium Term and Long Term Forecasts', *Geneva,* WMO/TD-No.630, pp. 278-280

Mason, S.J. (1998), 'Seasonal Forecasting of South African Rainfall using a Non-linear Discriminant Analysis Model', *International Journal Climatology*, vol. 18, pp. 147-164.

Mason, S.J., Joubert, A.M., Cosijn, C. and Crimp, S.J. (1996), 'Review of Seasonal Forecasting Techniques and their Applicability to Southern Africa', *Water SA*, vol. 22, pp. 203-209.

Mjelde, J.W., Thompson, T.N., Nixon, C.J. and Lamb, P.J. (1997), 'Utilising a Farm-level Decision Model to Help Prioritise Future Climate Prediction Research Needs', *Meteorology Applied,* vol. 4, pp. 161-170.

Murphy, A. (1994), 'Accessing the Economic Value of Weather Forecasts: An Overview of Methods, Results and Issues', *Meteorology Applied*, vol. 1, pp. 69-73.

National Department of Agriculture, (1997), *A Profile of Agriculture in South Africa,* Government Printers, Pretoria.

Schulze, G.S. (1989), 'Seisoensvariasies van reenval oor die somerreenstreke van Suid-Afrika' M.Sc. dissertation, University of Pretoria.

Simon, H.A. (1984), 'On the Behavioural and Rational Foundations of Economic Dynamics', *Journal of Economic Behavioural Organisation*, vol. 5, pp. 35-55.

Sonka, S.T. (1986), 'Information Management in Farm Production', *Computers and Electronics in Agriculture,* vol. 1, pp. 75-85.

Sonka, S.T., Changnon, S.A. and Hofing, S. (1988), 'Assessing Climate Information Use in Agribusiness. Part II: Decision Experiments to Estimate Economic Value', *Journal of Climatology*, vol. 1, no. 8, pp. 766-774.

Taljaard, J.J. (1986), 'Change of Rainfall Distribution and Circulation Patterns over Southern Africa in Summer', *Journal of Climatology,* vol. 6, pp. 579-592

Tennant, W.J. (1999), Numerical Forecasting of Monthly Climate in Southern Africa. *International Journal of Climatology*, vol. 19, pp. 1319-1336.

Tyson, P.D. (1986), *Climate Change and Variability in Southern Africa.* Oxford University Press, Cape Town.

van Heerden, J., Terblanche, D.E., and Schulze, G.C. (1988), 'The Southern Oscillation and South African Summer Rainfall', *Journal of Climatology,* vol. 8, pp. 577-597.

Vogel, C.H. (2000), 'Usable Science: An Assessment of Long-term Seasonal Forecasts amongst Farmers in Rural Areas of South Africa', *South African Geographical Journal,* vol. 82, pp. 107-116.

11 Meeting User Needs for Climate Forecasts in Malawi

NEIL WARD AND JOLAMU NKHOKWE

Effective climate information and prediction services require an appropriate framework where users recognize what is possible to predict, where the providers recognize what is essential to be predicted, and where the scientific information flow is in a form that can be readily assimilated in decision-making.

(Glantz, 2001, p. 22)

Introduction and Background

A growing interest in the ability of science to provide information about the climate of the coming several months developed during the early 1990s. The interest reached a crescendo with the 1997/98 El Niño event, which attracted worldwide attention to climate extremes, the extent to which they were attributable to El Niño, and the extent to which they had been forecasted. One criticism of the information provided by the climate community is that, despite its potential, it often fails to meet user needs. Addressing this situation has thus become a priority within the forecasting community. The aim is to provide information that better matches user needs and is communicated in an understandable manner so that it can more effectively influence decisions and contribute to alleviating negative impacts of climate anomalies.

Various activities could contribute to improving the fit between forecasts and user needs. These might include efforts to expand forecast products and accompanying multi-media products to explain them, and ongoing dialogs with user groups. The long-lead forecast component of the Malawi Environmental Management Project has enabled consideration of how to better meet user needs in Malawi, and provided opportunity to make progress. This chapter mainly draws on some of the results of that project.

First, we review the format of large-scale seasonal climate forecasts currently available and present their interpretation at smaller scales. Next, we provide examples of the types of information required by users in Malawi. These examples emerged from a survey,[1] user workshops in Malawi, and subsequent interviews with decision-makers from a variety of sectors that explored in detail the types of products that were both needed and technically feasible. Finally, we discuss the types of forecast information that can be realistically generated in response to the user needs that were identified, and present some graphics as examples.

Seasonal Climate Forecasts

There is now an established scientific basis to forecast the likelihood of the different climate scenarios expected across a specified region for the coming season (Palmer and Anderson 1994; Hastenrath 1995; Goddard et al., 2001). However, the potential is greater for certain regions and certain times of the year. For southern Africa, research has clearly demonstrated predictability for the rainy season, which for much of the region, occurs between November and March (Jury, Mulenga, and Mason, 1999; Mason and Tyson, 2001; Landman et al., 2001). Running dynamical and statistical climate models to produce forecasts for a long series of past years (typically 30-50 years) allows forecasters to quantify the approximate degree of reliability to expect of the forecasts, such that real-time forecasts can be reliably expressed in terms of the likelihood of different outcomes.

The consensus forecasts issued by the Southern Africa Regional Climate Outlook Forum (SARCOF) estimate the likelihood (probability) of three different rainfall categories for different regions across Southern Africa (Figure 11.1). The three categories (below-normal, near-normal and above-normal) are defined such that they each occur 10 times in a specified 30-year period. In other words, over a large number of years, they are each expected to occur 33.3% of the time.

The tercile concept is illustrated in Figure 11.2 for the Bvumbwe station in Malawi. In the figure, the January-March rainfall total for each year between 1961 and 1990 has been ranked from the wettest to the driest, and the year number is indicated at the top of each bar. For this location, the graph shows the boundaries of the three rainfall categories (terciles) that are used to express the SARCOF seasonal rainfall forecast. The boundary between the above-normal tercile and near-normal tercile lies between the 10th and 11th wettest seasons. The boundary between the two terciles can be calculated by averaging the 10th and 11th wettest seasons (1972 and 1988), here the average of 685.8mm and 681.5 mm, which is 683.6 mm. Consider a seasonal rainfall forecast for January-March. For Bvumbwe, a forecast of 60% likelihood of the above-normal rainfall tercile implies that there is a 6 in 10 chance that the January-March rainfall total will be above 683.6 mm. The boundary between the near-normal and below-normal terciles can be estimated by averaging the 10th and 11th driest seasons, which amounts to 509.6mm. Thus, a 30% probability forecast for the near-normal category implies a 3 in 10 chance of rainfall in the range 509.6mm to 683.6 mm. A 10% probability forecast for the below-normal category implies a 1 in 10 chance of rainfall below 509.6 mm.

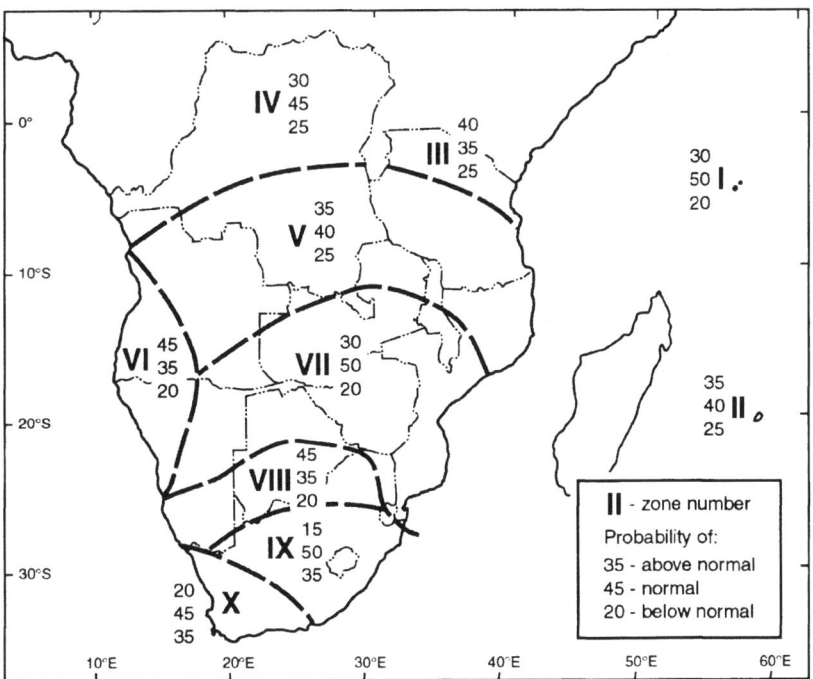

Source: Drought Monitoring Centre, 2002

Figure 11.1 Tercile forecast for southern Africa, January–March 2002

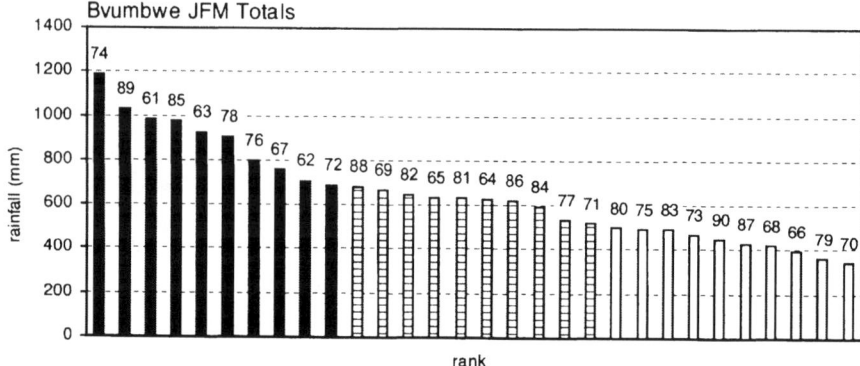

Figure 11.2   January–March rainfall total (mm) at Bvumbwe, Malawi, in each year 1961–90

This approach of estimating the likelihood of different categories (here, the three tercile categories) raises a number of key research issues. The tercile boundaries described above are estimated based on the distribution of actual rainfall values taken from a 30-year sample. More reliable estimates may be achieved by smoothing the distribution, or by quoting the boundaries to a more appropriate degree of accuracy (e.g. the nearest 10 mm). There are also questions regarding the spatial scale to which the probability forecast is applicable. The skill of seasonal forecasts has typically been evaluated at large spatial scales, and the probability forecasts are calibrated based on the skill of the forecasts at those large scales. Available evidence suggests that skill decreases if the forecast is verified over finer spatial scales (e.g. Mutai, Ward, and Colman, 1998), though this has still to be evaluated definitively, and furthermore, downscaling methods may subsequently enhance skill at the smaller scales.

This tercile approach raises a number of key research issues. The tercile boundaries described above are estimated based on the distribution of actual rainfall values taken from a 30-year sample. More reliable estimates may be achieved by smoothing the distribution, or by quoting the boundaries to a more appropriate degree of accuracy (e.g., the nearest 10 mm). There are also questions regarding the spatial scale to which the probability forecast is applicable. The skill of seasonal forecasts has typically been evaluated at large spatial scales, and the probability forecasts are calibrated based on the skill of the forecasts at those large scales.

Verifications have been performed assuming forecasts are applicable to rainfall averaged over about 50,000 km². Available evidence suggests that skill decreases if the forecast is verified over finer spatial scales (Mutai, Ward, and Coleman, 1998).

Climate Information Requested by Users

For many activities that are influenced by climate (e.g. related to agriculture, water resource management, road maintenance), a range of different types of climate information emerge as potentially useful. A general observation from the Malawi survey was that, given the option, most users would prefer a package composed of various types of climate information tailored to suit their needs. Nonetheless, most users find at least some value in general information products, such as the long-lead seasonal climate outlook or the 10-day agrometeorological monitoring bulletin created by Malawi Meteorological Department. One challenge for meteorological services is to find the best balance between generic products targeted at broad audiences and tailored packages targeted at specific users. The latter add a significant amount of value, but at a considerable cost.

Although many users may find long-lead forecasts valuable, they also need other types of climate information to provide a more complete perspective on the range of climate variability to plan for. Furthermore, once the seasonal forecast has been made, most users can reap additional value from tailored monitoring of the climate anomalies that impact their activities. Indeed, providing such information can potentially enhance the average value of the long-lead forecast, by increasing the confidence in a forecast and by alerting a user to updated information about the season should it be evolving in a manner that was not given a high likelihood by the forecast. Generating such information in objective ways is feasible and could be a productive topic for research and development. Links can also be drawn between climate monitoring information and weather forecast information. There is clearly potential to integrate long-lead climate forecasts with a range of additional meteorological and climate information.

Some examples of the types of information requested by users or potential users are described below. The different types of information requested include information about the potential evolution of the season, downscaled forecasts (both spatially, and in terms of the character of weather events through the season), seasonal monitoring and historical climate information. While rainfall is a key variable of interest, many other variables were also requested such as sunshine hours and water balance. Each user group tended to identify unique needs requiring tailored packages that integrate different types of information. Feedback from the survey (which included agriculture extension services) and associated user workshops in Malawi, recognized that the needs of small-holder farmers represented a critical, and challenging, area for subsequent more detailed evaluation. The users referred to below were identified based on their response to the user survey and were subsequently interviewed to explore their needs in more detail.

One NGO concerned with food security and children's health was eager to receive climate information that might indicate upcoming risks to food security. Recognizing climate to be one determining factor in food security, the NGO identified some specific product types that would enable them to integrate climate information effectively into evaluations of vulnerability. For this NGO, the need for graphics to communicate the range of possible outcomes was emphasized, along with the need to visualize the different potential ways that a season might evolve. Such a perspective on the potential evolution of the season was considered essential, since overly deterministic perceptions of forecasts can prove damaging.

A cumulative rainfall graphic that communicates the range (ensemble) of possible outcomes expected in the coming season for particular locations emerged as likely to make a valuable contribution to meeting this user's needs. Such a graphic could be complemented with a similarly constructed product that monitors the 10-day rainfall evolution throughout the season for the areas that the NGO requests. By comparing the season's evolution with a range of past years, the NGO could evaluate the extent to which the current season is evolving in an anomalous way, relative to the long-term climate and the long-lead climate forecast.

An association that represents and provides information to a number of agriculture estates can serve as a vehicle for disseminating tailored climate information to a large number of users. The association interviewed had many members, and interacted with these members at least once a week, so tailored weather and climate information could be provided relatively frequently. Two key areas where climate information could be useful were identified by the association: first, regarding decisions about irrigation, which are made throughout the season; and second, regarding fertilizer, which is applied at the onset of the season, and again in mid-season.

For strategy decisions made at the outset of the season, downscaled seasonal forecast information was requested. The SARCOF product, interpreted in terms of the probability of precipitation ranges for specific estate locations, would provide a good starting point. For application of the fertilizer itself, the coupling of climate monitoring and weather forecasting was requested. Monitoring was requested not just in terms of 10-day rainfall totals, but also in terms of the evolution of the water balance at estate locations. This is to evaluate the suitability of the ground for fertilizer, and to guide choice of fertilizer. The generation of such information, by merging satellite and *in situ* observations, could be another productive area for research. Weather forecasts, together with the water balance monitoring, could then lead to the best strategy for fertilizer application.

Regional water boards manage water supply in Malawi. With multiple water sources, they must make complex decisions regarding which source to utilize and to what extent, given that there are different costs attached to them. For example, the board faced a decision over how to best utilize one source that required substantially more energy to pump the water. A package of information including long-lead forecasted rainfall, monitored rainfall conditions, and historical rainfall distributions were requested for stations in the vicinity of the reservoirs. There would be potential value in quantitatively estimating the expected inflow to reservoirs given the long-lead forecast, and the rainfall conditions of the last 10 days. The first step could again be the SARCOF product interpreted in terms of probabilities of the tercile ranges in millimeters. For monitoring, a graphic was envisaged with the normal cumulative rainfall evolution of the season, with the evolving season superimposed, updated at a frequency of every 10 days. Such estimates require more substantial research and development initiatives.

Tourism makes a substantial contribution to the Malawi economy. Discussions with a tourist company revealed a wide range of weather and climate information that would help them operate more efficiently and deliver improved service to tourists. A long-lead forecast would promote better management of tourism in both Game Reserve Parks and Lake Shore locations. One of the key decisions for tour operators is the balance between promoting and running Game Reserve Park tourism and tourism on the Lake Shore. When the Game Reserve Parks become too wet, roads can become difficult and Lake Shore tourism becomes more attractive. The seasonal forecast could thus be useful for developing a seasonal strategy, whereas 10-day rainfall monitoring would be useful for short term planning, such as avoiding likely impassable roads in the Game Reserves. Here, coupling of the seasonal forecasts, shorter-term weather forecasts, and local monitoring of conditions would again provide maximum information. Additional information, such as the number of sunshine hours, was of interest, thus graphics of the sunshine recorded at lake locations over the previous days could supplement the forecast information provided.

Responding to User Needs

The enhancement of climate information is evolving amidst a background of rapidly developing computing and communication technology coupled to significant advances in climate science. This is increasing the potential for tailored climate information to be generated by experts in Malawi with personal computers and delivered to users via electronic communication. Developments in science and technology are also enhancing the potential for climate products such as satellite images of precipitation and the land surface.

Several key areas emerged in the survey that, if addressed, could make forecasts more useful. These included such aspects as interpretation of the forecasts, communication, and provision of additional, supplementary information to the forecasts. Many participants in the survey requested an interpretation of the long-lead forecast for their geographical area of activity. Follow-up interviews explored possibilities for producing and communicating such information.

It was found that, even with a large-scale tercile forecast, valuable information can be added by interpreting and communicating the meaning of the forecast for individual locations and sub-areas. Figure 11.2 illustrated one way to visualize the meaning of the forecast for a particular location (Bvumbwe, Malawi), while at the same time providing a perspective on the historical rainfall amounts. Figure 11.3 provides a similar graphic for Nkhotakota, located in the north of Malawi on the Lake. This area tends on average to receive more rainfall than Bvumbwe, which is located in the south. For Nkhotakota, the boundary between near-normal and above-normal rainfall categories is much higher (1035.0 mm). Thus, for this location, a 60% probability forecast for the above-normal category implies a 6 in 10 chance that rainfall will be above the estimated tercile boundary of 1035.0 mm.

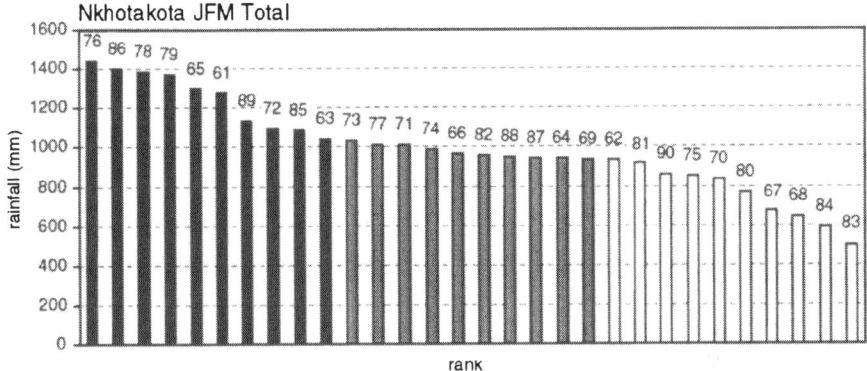

Figure 11.3 January–March rainfall total (mm) at Nkhotakota, Malawi, in each year 1961–90

These interpretations can be viewed as very simple forms of statistical downscaling. Although the forecast for these two locations is very different in terms of amount of rainfall, the forecast is the same in terms of ranked departure from normal (i.e., in terms of the probability of the different tercile categories occurring). One of the current research challenges is to identify if and how spatial variations in rainfall anomalies can be forecast over scales of less than a few hundred kilometers.

Both the agriculture and water resource sectors were interested in the possible patterns of seasonal evolution of the climate. Such a perspective adds greatly to the way users perceive the upcoming season and its relation to the forecast. Cumulative rainfall graphs for past years are one way to communicate the range of seasonal evolutions to plan for. For example, the December to March pattern is shown in Figure 11.4 for ten years, evenly sampled from a ranked listing of the 30 years in the database (see Figure 11.2). This gives an indication of the range of daily rainfall sequences to plan for, if the coming season is considered to have no tendency toward being wetter or drier than normal (i.e. in the absence of a seasonal forecast that alters the climatological probability distribution). The tick marks on the right hand axis of Figure 11.4 indicate the boundaries between the rainfall tercile categories.

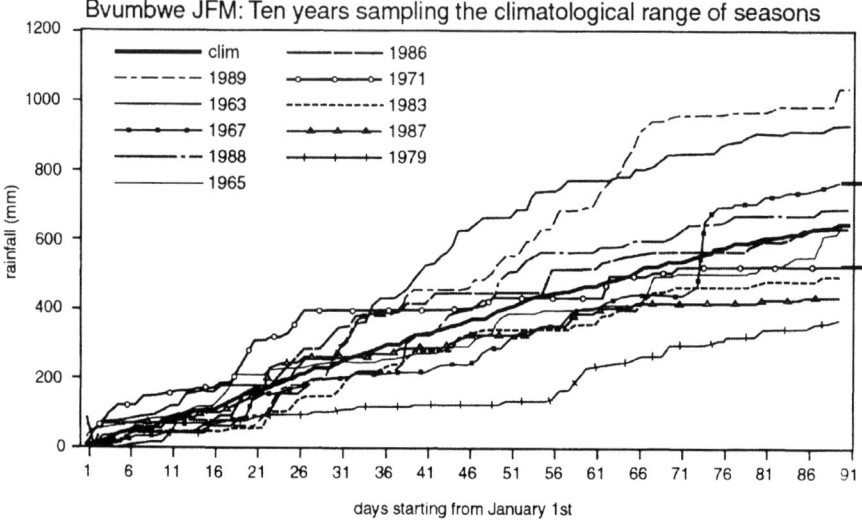

Figure 11.4   A typical set of daily rainfall evolutions of the January–March season at Bvumbwe, Malawi

This provides a sample of the range of possible patterns to plan for, even in the absence of a seasonal forecast. However, if a forecast is available, for example stating a probability of 60% for above-normal conditions, 30% for near-normal conditions, and 10% for below-normal conditions, it is possible to sample six years from the above-normal set, three years from the near-normal set, and one year from the below-normal set and plot a similar graph. In Figure 11.5, the years have been selected such that the graph provides an indication of the range of daily rainfall sequences to plan for, given the forecast. The resulting 10 years thus illustrate the range of seasonal evolutions that the forecast implies. Such charts can also be used as a basis for monitoring the season, for example, by updating the observed cumulative rainfall at a frequency of every 10 days. This immediately provides the user with a perspective on how unusual the evolution of the current season is in relation to past seasons, and in relation to the probability forecast.

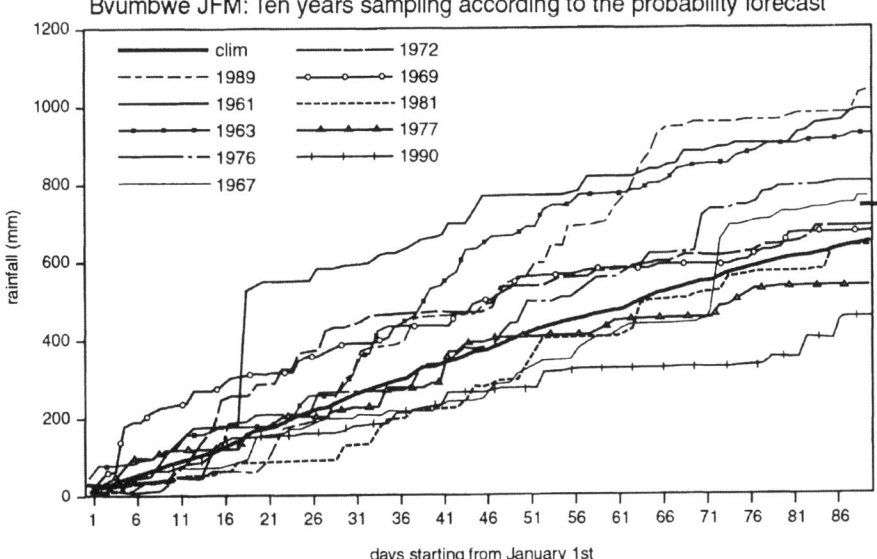

Figure 11.5 **Daily rainfall evolutions of the January–March season at Bvumbwe, Malawi, where the ten years have been sampled according to the probability forecast (6 yrs above normal, 3 normal, 1 below normal)**

The users interviewed for the study expressed an interest in additional information to complement the seasonal forecasts and other products. For agricultural users, monitoring and long-lead forecasting of the water balance at specific sites was requested. By driving a simple water balance model with meteorological observations, it is possible to provide such additional information. Interpretation of the long-lead SARCOF product could be made in terms of water balance by calculating the water balance characteristics for each of the years in the historical set. Indeed, each of the bar charts for rainfall in Figures 11.4 and 11.5 could be replaced by the water balance for the same years, to arrive at the distribution of water balances to expect conditional upon the observed rainfall tercile. Streamflow or inflow into specified reservoirs could also be treated in this way. However, arriving at such forecasts via the tercile rainfall forecast is likely not the optimal approach, and research into the creation of forecast information for the variables needed is a key research area.

Conclusions

Within the climate scientific community, a great emphasis has been placed on developing the ability to forecast large-scale rainfall patterns. However, users often requested tailored packages that integrate a variety of information, including more detailed features of the expected rainfall, other climate variables, and information about the consequences of the expected climate. As illustrated in this chapter, a number of simple methods of statistical downscaling can transform the SARCOF product into the type of rainfall information requested by many users surveyed in Malawi. This type of information can be effectively coupled with monitoring products and other information, including weather forecasts, to form a package of tailored climate information. Indeed, tailored products appear to have a place alongside more generic information products, such as the 10-day agrometeorological bulletin produced by Malawi Meteorological Department and global or regional scale products, such as satellite precipitation estimates and regional climate outlook forum products.

While simple statistical downscaling can provide approximations to some of the tailored information requested by users, much better products can be envisaged through research and development of more sophisticated downscaling tools. Such research requires collaboration across disciplines and across research and operational institutions. The survey and subsequent interviews suggest that a large number of users need the extra value that can be gained from a tailored package.

In Malawi, the process continues towards finding the best way to identify and provide the information needed by users. A National Meteorological Committee comprised of meteorologists, users, and policy makers has been established to address these issues. The results from the study presented here thus represent an early step in a longer process of linking climate science and society's needs.

Acknowledgments

The authors are grateful to Mr. H. Dandaula (former Director of Malawi Meteorological Department) and Mr. D. Kamdonyo (current Director) for their advice and support through the execution of this work. The Malawi

Environmental Management Project, supported by the World Bank, provided financial support. The input and advice of the Social Science Department, University of Malawi, is greatly appreciated, especially for their lead in constructing and implementing the user needs survey.

Note

1   Over 100 potential forecast users responded to the written questionnaire. This survey was implemented as a collaboration between the Malawi Meteorological Department and the Centre for Social Research at the University of Malawi. The results of this survey are described in a detailed report issued by the Malawi Meteorological Department (Mkandawire, Brouder, and Ward, 1999).

References

Drought Monitoring Centre (2002), Southern Africa Regional Climate Outlook Forum (SARCOF), http://www.dmc.co.zw/.

Glantz, M. (2001), *Lessons Learned from the 1997-98 El Niño: Once Burned, Twice Shy,* A UNEP/NCAR/UNU/WMO/ISDR Assessment, October 2000.

Goddard, L., Mason, S.J., Zebiak, S.E., Ropelewski, C.P., Basher, R., and Cane, M.A. (2001), 'Current Approaches to Seasonal-to-Interannual Climate Predictions', *International Journal of Climatology*, vol. 21, pp. 1111-1152.

Hastenrath, S. (1995), 'Recent Advances in Tropical Climate Prediction', *Journal of Climate,* vol. 8, pp. 1519-1532.

Jury, M. R., Mulenga, H.M., and Mason, S.J. (1999), 'Development of Statistical Long-range Models to Predict Summer Climate Variability over Southern Africa', *Journal of Climate*, vol. 12, pp. 1892-1899.

Landman, W. A., Mason, S.J., Tyson, P.D., and Tennant, W.J. (2001), 'Retroactive Skill of Multi-tiered Forecasts of Summer Rainfall over Southern Africa', *International Journal of Climatology*, vol. 21, pp. 1-19.

Mason, S. J., and Tyson, P.D. (2001), 'The Occurrence and Predictability of Drought over Southern Africa' in D.A. Wilhite, (ed.), *Drought. Volume 1: A Global Assessment.* Routledge, New York, pp. 113-134.

Mkandawire, E., Brouder, A. and Ward, M.N. (1999), *Climate Information User Needs Assessment Survey in Malawi,* Malawi Meteorological Department, unpublished report.

Mutai, CC., Ward, M.N., and Colman, A.W. (1998), 'Towards the Prediction of the East Africa Short Rains based on SST-atmosphere Coupling', *International Journal of Climatology*, vol. 9, pp. 975-997.

Palmer, T.N. and Anderson, D.L.T (1994), 'The Prospects for Seasonal Forecasting–A Review Paper', *Quarterly Journal of the Royal Meteorological Society*, vol. 120, pp. 755-793.

12 A Future for Forecasts?

KAREN O'BRIEN AND COLEEN VOGEL

> The weather this season is not the same as it was a year ago, and common experience leads us to suspect that it will be different still a year hence. None of us, including the experts in long-range forecasting, have a reliable idea of how it will differ.
>
> (Cane, 2000, p. 29)

Introduction

Each year, in one part of the world or another, climate extremes focus substantial attention on communities and places at risk to environmental change. Such is the growing concern that several scientific networks are expending great time and resources to assess how the science of global environmental change can be improved, and what can be done to enhance resilience in the face of anticipated changes (e.g., World Resources Institute, 2000; McCarthy et al., 2001; IGBP, 2001). Most of these groups acknowledge that we are living at a time of risk and change that requires collective action.

Projected climate changes during the 21st Century, for example, may potentially lead to future large-scale and possibly irreversible changes in the Earth system, resulting in differential impacts at global, national, and sub-national scales (McCarthy et al., 2001). Temperature, precipitation, and other variables are likely to change, with impacts on soil moisture, water availability, pests, etc. In the absence of comprehensive climate change abatement policies, attention has focused increasingly on adaptation strategies aimed at mitigating negative impacts, reducing vulnerability, and increasing resilience. However, because climate change will affect both the mean and variability of different climate parameters, adaptation strategies must focus on addressing both stresses and shocks.[1]

One of the ways to potentially enhance the capacity to cope with environment stresses and shocks may be through the effective use of seasonal climate forecasts. As we have seen in various chapters in this book, the science of forecasting has advanced to a level where seasonal forecasts can be used with some reliability in certain areas. Indeed, when coupled with other types of climate information, forecasts can contribute to decision-making in a variety of sectors.

In this chapter, we summarize some of the key issues surrounding forecast use, including their potential and limitations. We argue that information alone is insufficient for developing successful strategies for coping with climate variability in southern Africa, and we call for a widening of the discourse on seasonal forecasts to include the dynamic context that shapes forecast use in this region. To realize the potential value of seasonal climate forecasts in southern Africa requires policies and actions that enable different types of users to respond appropriately to climate variability and long-term climate change.

Forecasts: A Promising Tool

Seasonal climate forecasts represent a potentially powerful link between scientific research and applications. There is widespread optimism that the development and dissemination of climate forecasts can provide much-needed information to a variety of users, which will inevitably reduce losses and damages attributed to climate variability. Both in the context of famine early warning systems and seasonal climate outlooks, there is a belief that information is the missing link between adverse weather and improved coping strategies (Buchanan-Smith, Davies, and Petty, 1994).

According to this view, if information related to seasonal forecasts can be properly disseminated and acted upon, the information can contribute to economic gains as well as improved food security (Betsill, Glantz, and Crandall, 1997). Indeed, a study of the value of improved forecasts to U.S. agriculture showed that a reliable ENSO prediction could be worth an estimated USD 323 million (Solow et al., 1998). Countries such as Peru and Brazil also offer examples of how policymakers have successfully used forecast information to manage socio-economic responses to climate variability (Golnaraghi and Kaul, 1995).

Examining responses to climate forecasts in extreme years can serve as a means for improving the understanding of how different sectors cope with climate variability. For example, the strong El Niño that struck in mid-1997 unleashed a variety of impacts around the globe, with some figures estimating costs between USD 32 billion to USD 96 billion (Glantz, 2001).[2] This event, which some refer to as the "El Niño of the 20th Century," was the first case that saw the participation of numerous forecasting groups in real-time prediction and whose impact was forecasted accurately by meteorological centers well in advance (Barnston et al., 1994; Landsea and Knaff, 2000). From the perspective of climate scientists, many acknowledge that most of the dynamical models used to predict the ENSO of 1997/98 did a credible job. Nevertheless, several models underestimated the magnitude of the event, revealing an area where forecast skill can be improved: "The forecasts are far from perfect, especially so for the connections to local conditions with the greatest human consequences" (Cane, 2000, p. 49). From a societal perspective, the forecasts issued in advance of the 1997/98 El Niño offered an opportunity to prepare for the event, rather than simply react to it. As Dilley pointed out in Chapter 2, this marks a major departure from normal procedures in the food aid community.

Climate forecasts represent a potentially valuable tool for a wide range of users. In their current format, however, the value of forecasts is constrained by issues related to dissemination, access, and relevance, as highlighted in several chapters in this book. Many of the groups that are vulnerable to climate variability do not receive the forecasts, do not understand them, or cannot make substantive changes based on the information. In fact, many are excluded from the forecasting process, and their voices remain unheard in the discourse on climate forecasts. In the next two sections, we summarize some of the issues that undermine the value of seasonal forecasts to end users in southern Africa. These limitations make it clear that there is a wide gap between actual and potential benefits.

A Tool for Whom?

Scientists have yet to fully develop effective methods for disseminating probabilistic forecasts to end users, and the channels that do exist are imperfect. If information is power, then those institutions and individuals controlling the dissemination of climate forecasts are in a position to influence how and whether the message reaches end-users. Those at the receiving end of forecasts are generally assumed to be "empowered" by the information that they receive, in that that they can take actions that will improve their well-being. Yet the case studies on forecast use presented in this book and elsewhere (e.g., Nelson and Finan, 2000; Roncoli, Ingram, and Kirshen, 2000, 2001; Broad and Agrawala, 2001) show that information dissemination and communication is not a neutral process, and that forecasts alone are not necessarily empowering.

Information dissemination requires some kind of network, whether through radio, farmer unions, or extension services, where information flows both from the center outwards, and from the grass roots level upwards. The choice of dissemination networks by forecast producers and government bodies leads to the intentional or unintentional exclusion of some groups from receiving the information. For this reason, Blench cautions in Chapter 4 against considering only "official networks" for the dissemination of forecasts, arguing for the use of unofficial channels of information flow that are more relevant to small-scale farmers. As Cash (2000) argues, forecasts should ideally be located within a system or series of "support systems" and networks that are understood by all role players in the system or network.

Dissemination alone does not determine who receives the forecasts. Access to climate information is also determined by the local context, including internal dynamics and "politics" within organizations such as government ministries or even households. As Phillips points out in Chapter 7, some users have greater access to forecasts than others. Factors such as village politics, ethnicity, and gender influence this access (Roncoli, Ingram, and Kirshen, 2001). Recognizing that access to forecasts is unequal is a prerequisite for understanding and improving user responses.

Even if dissemination and access were not issues, there are concerns that the information provided by forecasts is not valuable to many potential end users because of its coarse spatial resolution and because it provides no details on the onset and timing of rainfall. Although small-scale farmers, commercial farmers, and agribusinesses alike place a high value on forecast information, the information that is currently included in seasonal forecasts is not detailed enough to have any real impact on management strategies. The case studies of Hudson and Vogel (Chapter 5), Bohn (Chapter 6), Klopper and Bartmann (Chapter 10), and Ward and Nkhokwe (Chapter 11) showed that information needs vary among different types of users, and that the forecasts should be targeted to meet the needs of a variety of different user groups. As Kihupi and his coauthors emphasize in Chapter 8, many farmers already make widespread use of local indicators for seasonal climate predictions, and seasonal forecasts represent one more piece of information that they incorporate into their decision-making process.

The production of forecasts and the target audience for whom they are intended needs to be examined in greater detail. Are climate forums, such as SARCOF, only focused on getting forecasts to "large" players who then have responsibility for ensuring the message "trickles down" to a variety of users? As Bohn demonstrates in Chapter 6, agribusinesses in Swaziland can potentially make use of forecasts at a number of points in the production process. In Chapter 10, Arndt, Bacou, and Cruz show that the agricultural marketing system in Mozambique could make good use of seasonal forecasts. Yet other case studies presented in this book reveal that a large number of "smaller-scale" end-users have expressed interest in being able to consider forecasts in their decision-making. These users, however, are often constrained in their ability to respond to seasonal forecasts by several factors, including access to resources and difficulties interpreting probabilistic forecasts. The question then arises: Is information sufficient?

Is Information Sufficient?

The creation and dissemination of climate forecasts involves a flow of information that can potentially have far-reaching repercussions. In modern economic theory, information is considered a factor in the decision process that can be utilized by potential users or decision makers to reduce uncertainty (Johnson and Holt, 1997). In terms of the market value of forecasts, "[a]dditional information should, in general, improve resource allocation and enhance market efficiency" (Johnson and Holt, 1997, p. 85). However, as Wilks (1997, p. 110) notes, "for the forecasts to have value, actions must be available that are capable of producing changes in the consequences. Otherwise the forecasts will offer no more than 'entertainment value'."

In theory, a wide range of benefits can accrue to users of seasonal climate forecasts, including increased crop production, reduced crop losses, or strategic marketing of livestock. Actions in response to forecasts may involve changing crops or seed varieties, sowing seeds earlier or later, placing more or less land under production, investment or disinvestments in irrigation or fertilizers, and a host of other strategies (Stern and Easterling, 1999). Such actions are referred to by Stern and Easterling (1999) as *ex ante* coping strategies that aim to reduce adverse impacts of climate events on agricultural output and profits, or enable farmers to exploit opportunities if conditions are favorable.

These strategies, however, vary among different end users. The ability to cope and adapt is often constrained by such factors including, in the case of small-scale farmers, the ability to gain access to resources that would assist them in their agricultural activities (e.g., land, labor, fertilizers, credit, etc.). The chapters in this book contribute to an accumulating body of evidence showing that the use of forecasts among small-scale or communal farmers in rural areas is constrained by the limited production alternatives available upon relatively short notice (Eakin, 2000; O'Brien et al. 2000; Roncoli, Ingram, and Kirshen, 2001).

Downing et al. (1997) observe that subsistence farmers are less likely to have the resources to consider anticipatory action, as compared to large-scale commercial farmers who are more likely to be linked to national markets and international agribusinesses, and who have a greater ability to invest in agricultural technology. Hulme (1994) argues that the use of the forecast information in rural communities is severely limited in part due to the limited flexibility of such systems to respond to external information. This is exacerbated by a lack of infrastructure to support producers' choices, in the rare cases where such choices can be made (Hulme, 1994).

Eakin (2000) indeed found that the ability to act upon climate information is limited among small-scale farmers in Tlaxcala, Mexico. Response strategies often require access to fertilizers, alternative crops, hybrid seeds, and commercial markets. These in turn are contingent upon the availability of credit, insurance, technical support, and market conditions. Eakin argues that political/economic uncertainties are more important than climate risk in determining production choices, and concludes that in their present form, "climate forecasts may be a technology poorly suited to the needs and constraints of small-scale farming in Mexico" (Eakin, 2000, p. 34).

In Northeast Brazil, Nelson and Finan (2000) found that farmers were unable to use the forecasts because they have limited choices of technology. It is not producer ignorance over what and when to plant that results in drought impacts, but the lack of alternatives to buffer livelihoods against extreme events. The Cearense Foundation for Meteorology and

Water Resources (FUNCEME) disseminated the forecasts with great fanfare and little attention to the constraints of the wider agricultural environment (e.g., access to credit, alternative technologies, and other inputs).

Roncoli, Ingram, and Kirshen (2001) and Vogel (2000) came to similar conclusions in their work in West Africa and South Africa. Farmers in Burkina Faso, for example, while keen to use forecasts, were clear to indicate that their use of forecasts was and is contingent on the availability of credit supplies and other inputs (Roncoli, Ingram and Kirshen, 2001). Small-scale, emerging farmers in the North West Province in South Africa also indicated that credit and ownership of land were factors that inhibited their potential use of forecasts (Vogel, 2000). O'Brien et al. (2000) found that information was not enough to influence coping strategies among small scale farmers in Namibia and Tanzania. Although only a small percentage of farmers actually heard the forecasts in 1997/98, among those that did, only limited actions could be taken. Small-scale farmers in particular lacked access to agricultural credit, fertilizers, alternative seeds, tractors, and other factors that are indispensable for responding to climate variability.

In practice, the use of seasonal climate forecasts has therefore not advanced to the point where the benefits have been fully exploited. At present, the link between information and action among end users is unevenly developed. Within the farming sector, as illustrated above, there are also wide differences in the ability to act upon forecast information, assuming that such information is evenly disseminated in the first place.

Widening the Discourse

As climate science develops and more skillful seasonal forecasts become available, it is likely that forecast dissemination and access will improve as well. To capture the full potential of seasonal forecasts, however, the discourse on climate forecasts must be widened beyond issues of accuracy and access. Climate forecasts must be discussed in the context of coping strategies and adaptation to climate variability and change. Within this context, the socio-economic and political factors that shape and enable coping and adaptation must be addressed.

Blench and Marriage (1998, p. 7) argue that the fields of climate forecasts (and climate change more broadly) are "metamorphosing from the technical to the socio-political." They suggest that the consequences of climate variability must be decoded as much for their political significance as for their predictive element. Forecasts are not neutral pieces of information, but instead carry political messages regarding expectations for the upcoming season.

Given the latitude of interpretation of probability-based forecasts, politics can affect agencies' interpretations and influence whether governments listen to and act upon a forecast. For example, in 1997/98, different Peruvian fishing companies tried to convince decision makers of the "right" view to adopt regarding the intensity and duration of the ENSO event and its implications for fish stocks (Pfaff, Broad, and Glantz, 1999). Competition between various institutions places pressure on employees to produce clear answers, sometimes resulting in the removal of caveats from the forecasts that are used and conveyed in press releases. The media may also sensationalize the information contained in the forecasts and distort the message. In Northeast Brazil, the politicization of climate forecasts raised expectations of impending solutions to the drought. When these failed to materialize, FUNCEME lost much of its credibility with small-scale farmers (Finan, 1999). These examples confirm that information is not "value free" and can be distorted to serve varying needs.[3]

The discourse on climate forecasts must be broadened to include the dynamic context within which forecasts are received. As Stern and Easterling (1999, pp. 136-137) note, the consequences of climate variability depend on more than climate itself, and may include factors such as "population growth in and migration to areas that experience large climate variations; economic and infrastructural development in such areas; the level of dependence of human populations on food and other essential goods and services delivered from outside their immediate vicinity..." As Thompson illustrates in Chapter 3, the flexibility (or inflexibility) to respond to climate forecasts is dynamically determined by policies and processes occurring at different scales.

Southern Africa is currently undergoing dramatic economic changes as a result of globalization of economic activity and continuing implementation of national-level structural adjustment programs (Leichenko and O'Brien, 2002). Export-led growth is increasingly regarded as a desirable strategy to promote growth and development throughout

Africa, and southern African countries are making significant efforts to promote international trade and encourage foreign investment (Collier, Greenaway, and Gunning, 1997; Onafowora and Owoye, 1998; Bigsten, 1999).

The effects of economic globalization are likely to be felt throughout the agricultural sector in southern Africa, through shifts in cropping patterns (favoring export crops), improved access to agricultural technologies such as improved seeds, and better access to credit for some farmers (DeRosa 1997; Aboum-Ongaro 1999; Leichenko and O'Brien 2002). Although rising exports may increase incomes of small-scale farmers by fostering production and sales of cash crops in many regions (Jacques, 1997), increased involvement in the international economy may also leave small farmers vulnerable to fluctuating world prices and changing terms of trade (World Bank, 2000).

Together, global and national economic changes may have a substantial impact on southern African farmers and their ability to cope with climate variability. A seasonal climate forecast projecting above normal rainfall might induce farmers to switch to or expand production of marketable crops. Market liberalization has, however, shifted marketing responsibility to farmers, who may lack marketing options if private companies do not find remote areas lucrative to service. Even if farmers are able to market surplus production, prices dictated increasingly by global markets may be so low as to make the additional effort unprofitable. A seasonal climate forecast for below-normal rainfall might encourage farmers to limit the area planted, or migrate to other areas in search of wage labor. Reducing the area under cultivation may increase reliance on commercial food sources, whose prices fluctuate depending on international supply and demand. At the same time, the market for agricultural wage labor is likely to contract when poor rainfall is projected by seasonal forecasts.

The *context for forecast use is thus as dynamic as the climate*, leading to different potential outcomes for any given forecast. User responses must be understood within this changing political and economic context.

Conclusions

Is there a future for forecasts in southern Africa? Inter- and intra-annual variability of rainfall may be considered the key climatic elements that determine the success of agriculture, particularly in the semi-arid tropical regions of the world (Sivakumar, 1998). In southern Africa, the availability of long-term, probability-based forecasts has important implications for agricultural production, and hence food security. It can also provide potentially valuable input into famine early warning systems, and contribute to more timely distribution of emergency food relief (Betsill Glantz, and Crandall, 1997, Dilley, Chapter 2). Responses to present-day climate variability form the cornerstone for adapting to future climate changes. In the anticipation of changes in the frequency and/or magnitude of extreme events associated with global climate change, there is indeed a need for improved seasonal forecasts (Joubert and Hewitson, 1997; Landsea and Knaff, 2000; Buizer, Foster and Lund, 2000; McCarthy et al. 2001).

Improved seasonal climate forecasts can thus potentially reduce vulnerability to climate hazards in Africa, which in combination with improved resource management can serve as an important adaptation strategy for addressing long-term climate change (Downing et al., 1997). Biological indicators, such as the flowering of vegetation or appearance of certain insects, have been traditionally used by farmers to forecast climate conditions for the upcoming season. However, these indicators are apparently becoming more and more unreliable as the global environment changes (Ingram, Roncoli, and Kirshen, 2002).

The socio-economic, political, and cultural dynamics in the region (e.g., HIV/AIDS, institutional capacity, economic crises, conflicts) suggest that one should pay serious attention to the factors that shape coping and adaptive capacities to a variety of shocks, including but not limited to climate variability. While issues such as institutionalization of forecasts, funding for forecast development, suitability of forecasts to small-scale users, and level of skill of forecasts are important (IRI, 2001), the daily realities that confront this region mean that forecasts are not necessarily *the* solution to coping with climate variability.

There are other key issues that need to be integrated into forecast production and dissemination to ensure that they are actually used. Over the past decade, the continued development of seasonal forecasts has been justified based on both actual and potential value. While the actual value has been illustrated in numerous case studies and anecdotes, the potential value can only be speculated. To realize the potential value of seasonal climate forecasts will require a better understanding of the linkages between information needs and coping strategies. These linkages are likely to include policies and actions that enable different user groups to anticipate and prepare for climate extremes, rather than simply react to them.

Within the growing literature on climate forecasts, there is a tendency to identify and publicize the success stories, while the failures are seldom reflected upon, and often dismissed as "growing pains." The failures of forecasts to impact coping strategies can, however, offer valuable insights into the possibilities and limitations of climate forecasts. Understanding the role of various types of institutions in the process of forecast production and their uptake is also fundamental. Broad and Agrawala (2001) conclude that for forecasts to be useful, they must be socially robust, and the product should be the outcome of the interaction between data and other results, between people and environments and between applications and implications. The lack, moreover, of what Orlove and Tosteson (1999) call "institutional fit" can constrain forecast use and, in some cases, challenge institutional viability. The idea that "one forecast fits all" has been shown through case studies presented in this book and elsewhere to be unrealistic.

In the wake of the 1997/98 ENSO event, there is a need to critically reflect upon the potential benefits of seasonal climate forecasts. Case studies on user responses to seasonal forecasts illustrate a wide and heterogeneous range of responses, depending on the characteristics of the users and the context in which the forecasts are received. Case studies show that forecasts benefit different sectors and social groups, and that some groups are more prepared (or less prepared) to make use of the information. The anticipated return of El Niño conditions in the future necessitate that forecasts, as one tool for enhancing coping to extreme events, are not used exclusively as an "emergency" response mechanism, but that they begin to become used for longer-term risk-reduction and mitigation before an event occurs. Such preparation, however, cannot be undertaken concurrent with the forecast lead time. Appropriate structures and policies must be in place before the need arises. Coping with climate variability must therefore be considered a long-term priority, rather than a crisis to be managed on an *ad hoc* basis.

Climate forecasts have created opportunities as well as challenges for southern Africa. As this book has shown, the challenges relate to acknowledging the context and circumstances that influence the daily realities of end users. There is therefore a need to develop a parallel process to the production and dissemination of forecasts that focuses on better understanding of how end-users currently make management decisions, what factors constrain or enable them to make these decisions, and on the identification of critical factors that influence vulnerability and resilience to climate variability.

Notes

1   Environmental stresses can be described as negative pressures over a period of time, whereas shocks tend to arise unexpectedly, often with a more violent impact (De Haan, 2000). Drought provides an example of an environmental stress, whereas flooding represents an environmental shock. Both of these conditions are influenced by short-term climate variability and possibly long-term climate change.

2   Glantz (2001) gives an assessment of over 16 countries, their response to the event and the role that various mitigation activities played and could have played with respect to forecasts.

3   Climate forecasts alone cannot take credit for prompting actions or responses to extreme climate events. Often it is other indicators that stimulate responses to climate variability. In Ethiopia, the government became concerned about the food crisis of 2000, not because of seasonal forecasts, but because of the famine and the resulting media pressure to take action (Broad and Agrawala, 2001).

References

Aboum-Ongaro, W. (1999), 'Agriculture, Policy Impacts and the Road Ahead', in S.Kayizzi-Mugerwa (ed.), *The African Economy: Policy, Institutions and the Future*, Routledge, New York, pp. 248-263.

Barnston, A.G., van den Dool, H.M., Zebiak, S.E., Barnett, T.P., Ji, M., Rodenhuis, D.R., Cane, M.A., Leetmaa, A., Graham, N.E., Ropelewski, C.R., Kousky, V.E., O'Lenic, E.A., and Livezey, R.E., (1994), 'Long-lead Seasonal Forecasts – Where do we Stand?', *Bulletin of the American Meteorological Society*, vol. 75, pp. 2097-2114.

Betsill, M.M., Glantz, M.H. and Crandall, K., (1997), 'Preparing for El Niño: What Role for Forecasts?' *Environment*, vol. 39, no. 10, pp. 6-28.

Bigsten, A. (1999), 'Looking for African Tigers', in S. Kayizzi-Mugerwa (ed.), *The African Economy: Policy, Institutions and the Future*, Routledge, New York.

Blench, R. and Marriage, Z. (1998), 'Climatic Uncertainty and Natural Resource Policy: What Should the Role of Government Be?', Overseas Development Institute, *Natural Resource Perspectives*, vol. 31, pp. 1-9.

Broad, K. and Agrawala, S. (2001), 'The Ethiopia Food Crisis: Uses and Limits of Climate Forecasts, *Science*, vol. 289, pp. 1693-1694.

Buchanan-Smith, M., Davies, S. and Petty, C. (1994), 'Food Security: Let Them Eat Information', *IDS Bulletin*, vol. 25, no. 2, pp. 69-80.

Buizer, J., Foster, J. and Lund, D. (2000), 'Global Impacts and Regional Actions: Preparing for the 1997-98 El Niño', *Bulletin of the American Meteorological Society*, vol. 81, No. 9, pp. 2121-2139.

Cane, M.A., Zebiak, S.E. and Dolan, S.C. (1986), 'Experimental Forecasts of El Niño', *Nature*, vol. 321, pp. 827-832.

Cane, M. (2000), 'Understanding and Predicting the World's Climate System', in G.L. Hammer, G.L., N. Nicholls and C. Mitchell (eds), *Applications of Seasonal Climate Forecasting in Agricultural and Natural Ecosystems, The Australian Experience*, Kluwer Academic Publishers, Dordrecht, pp. 29-50.

Cash, D. (2000), 'Distributed Assessment Systems: An Emerging Paradigm of Research, Assessment and Decision Making for Environmental Change', *Global Environmental Change*, vol. 10, pp. 109-120.

Collier, P., Greenaway, D. and Gunning, J.W. (1997), 'Evaluating Trade Liberalization: A Methodological Framework', in A. Oyejide, I. Elbadawi, and P. Collier (eds), *Regional Integration and Trade Liberalization in Sub-Saharan Africa*, St. Martin's Press, New York.

De Haan, L.J. (2000), 'Globalization, Localization and Sustainable Livelihood', *Sociologia Ruralis*, vol. 40, no. 3, pp. 339-365.

DeRosa, D.A. (1997), 'Regional Integration and the Bias Against Agriculture and 'Disadvantaged' Sectors in Sub-Saharan Africa', in A. Oyejide, I. Elbadawi, and P. Collier (eds), *Regional Integration and Trade Liberalization in Sub-Saharan Africa*, St. Martin's Press, New York, pp. 256-305.

Downing, T.E., Ringius, L., Hulme, M. and Waughray, D. (1997), 'Adapting to Climate Change in Africa', *Mitigation and Adaptation Strategies for Global Change*, vol. 2, pp. 19-44.

Eakin, H. (2000), 'Smallholder Maize Production and Climatic Risk: A Case Study from Mexico', *Climatic Change*, vol. 45, pp. 19-36.

Finan, T.J. (1999), "Drought and Demagoguery: A Political Ecology of Climate Variability in Northeast Brazil", Paper presented to the Carnegie Council on Ethics and International Affairs, New York.

Glantz, M. (ed.) (2001), *Once Burned, Twice Shy? Lessons Learned from the 1997-98 El Niño,* United Nations University, New York.

Golnaraghi, M. and Kaul, R. (1995), 'Responding to ENSO: The Science of Policymaking', *Environment*, vol. 37, no. 1, pp. 16-44.

Hulme, M. (1994), 'Using Climate Information in Africa: Some Examples Related to Drought, Rainfall Forecasting and Global Warming', *IDS Bulletin*, vol. 25, no. 2, pp. 59-68.

IGBP (International Geosphere Biosphere Programme) (2001), Global Change and the Earth System: A Planet under Pressure, The Global Environmental Change Programmes, IGBP Science, No. 4., Stockholm.

Ingram, K., Roncoli, C. and Kirshen, P. (2002), 'Opportunities and Constraints for Farmers of West Africa to Use Seasonal Precipitation Forecasts with Burkina Faso as a Case Study', *Agricultural Systems*, in press.

International Research Institute for Climate Prediction (IRI) (2001), *Coping with the Climate: A Way Forward, Preparatory Report and Full Workshop Report*, a multi-stakeholder review of Regional Climate Outlook Forums concluded at an international workshop, 16-20 October 2000, Pretoria, South Africa.

Jacques, G. (ed.) (1997), *Structural Adjustment and the Poverty Principle in Africa*, Ashgate Publishing, Brookfield, VT.

Johnson, S.R. and Holt, M.T. (1997), 'The Value of Weather Information', in R.W. Katz and A.H. Murphy (eds), *Economic Value of Weather and Climate Forecasts*. Cambridge University Press, Cambridge, pp. 75-103.

Joubert, A. and Hewitson, B. (1997), 'Simulating Present and Future Climates of Southern Africa using General Circulation Models', *Progress in Physical Geography*, vol. 21, pp. 51-78.

Landsea, C.W. and Knaff, J.A. (2000), 'How Much Skill was there in Forecasting the Very Strong 1997-1998 El Niño?', *Bulletin of the American Meteorological Society*, vol. 81, pp. 2107-2119.

Leichenko, R.M. and O'Brien, K.L. (2002), 'The Dynamics of Rural Vulnerability to Global Change: The Case of Southern Africa', *Mitigation and Adaptation Strategies for Global Change*, vol. 7, pp. 1-18.

McCarthy, J.J., Canziani, O.F., Leary, N.A., Dokken, D.J., and White, K.S. (2001), *Climate Change 2001: Impacts, Adaptation, and Vulnerability*. Contribution of Working Group II to the Third Assessment Report of the Intergovernmental Panel on Climate Change. Cambridge University Press, Cambridge.

National Research Council. (1996), Learning to Predict Climate Variations Associated with El Niño and the Southern Oscillation: Accomplishments and Legacies of the TOGA Program, National Academy Press, Washington, DC.

Nelson, D.R. and Finan, T.J. (2000), 'The Emergence of a Climate Anthropology in Northeast Brazil', *Practicing Anthropology*, vol. 22, pp. 6-10.

NOAA Office of Global Programs (OGP). (1999), *An Experiment in the Application of Climate Forecasts: NOAA-OGP Activities Related to the 1997-98 El Niño Event*, Office of Global Programs, National Oceanic and Atmospheric Administration, U.S. Department of Commerce, Washington, DC.

O'Brien, K., Sygna, L., Naess, L.O., Kingamkono, R. and Hochobeb, B. (2000), *Is Information Enough? User Responses to Seasonal Climate Forecasts in Southern Africa*. CICERO Report 2000:3, Oslo, Norway.

Onafowora, O. and Owoye, O. (1998), 'Can Trade Liberalization Stimulate Economic Growth in Africa?', *World Development*, vol. 26, no. 3, pp. 497-506.

Orlove, B. and Tosteson, J. (1999), 'The Application of Seasonal to Interannual Climate Forecasts based on El Niño – Southern Oscillation (ENSO) Events: Lessons Learned from Australia, Brazil, Ethiopia, Peru and Zimbabwe. Working Papers in Environmental Policy, Institute of International Studies, University of California, Berkeley.

Pfaff, A., Broad, K. and Glantz, M. (1999), 'Who Benefits from Climate Forecasts?', *Nature*, vol. 397, pp. 645-646.

Roncoli, C, Ingram, K. and Kirshen, P. (2000), Can Farmers of Burkina Faso use Rainfall Forecasts?', *Practicing Anthropology*, vol. 22, pp. 24-28.

Roncoli, C., Ingram, K. and Kirshen, P. (2001), 'The Costs and Risks of Coping with Drought: Livelihood Impacts and Farmer's Responses in Burkina Faso', *Climate Research*, vol. 19, pp. 119-132.

Sivakumar, M.V.K. (1998), 'Climate Variability and Food Vulnerability', Global Change Newsletter (IGBP), vol. 35 (September 1998), pp. 14-17.

Solow, A.R, Adams, R.F., Bryant, K.J., Legler, D.M., O'Brien, J.J., McCarl, B.A., Nayda, W., and Weiher, R. (1998), 'The Value of Improved ENSO Prediction to U.S. Agriculture', *Climatic Change,* vol. 39, pp. 47-60.

Stern, P.C. and Easterling, W.E. (eds) (1999), *Making Climate Forecasts Matter.* National Academy Press, Washington, D.C.

Vogel, C.H. (2000), 'Usable Science: An Assessment of Long-term Seasonal Forecasts amongst Farmers in Rural Areas of South Africa', *South African Geographical Journal*, vol. 82, pp. 107-116.

Wilks, D.S. (1997), 'Forecast Value: Prescriptive Decision Studies', in R.W. Katz and A.H. Murphy (eds), *Economic Value of Weather and Climate Forecasts*, Cambridge University Press, Cambridge, pp. 109-146.

World Bank (2000), *Can Africa Claim the 21st Century?* The World Bank, Washington, DC.

World Resources Institute (2000), A Guide to World Resources 2000-2001: People and Ecosystems: The Fraying Web of Life, World Resources Institute, Washington.

Index